Automatic Cartography and Planning

Automatic Cartography and Planning

by the Experimental Cartography Unit,
Royal College of Art, London

This report is a feasibility study commissioned by the former Department of Economic Affairs on behalf of the East Anglia Economic Planning Council. It has been prepared by the Experimental Cartography Unit, Royal College of Art, London. The views and opinions expressed in the report are solely those of the Unit.

December 1969 (with addenda to September 1970)

Architectural Press, London

ISBN 0 85139 076 5

First published 1971

Text printed by Compton Printing Ltd., London and Aylesbury
Maps printed by Cook Hammond & Kell, Mitcham
Bound by BKT Printers Ltd., Tonbridge

Acknowledgements

To a substantial extent the results of this project have depended on the equipment and staff already assembled in the Experimental Cartography Unit; the funding and initiative for this have, in their turn, depended on the Natural Environmental Research Council and the Royal College of Art, who have agreed that the Unit should extend its activities into this field of "human" as opposed to "physical" geography.

Coppock, Prof. J. T. (Department of Geography, University of Edinburgh) – Consultant

Clayton, Prof. K. (School of Environmental Science, University of East Anglia)

Gaits, Mr. G. (Ministry of Housing and Local Government)

Rosing, Mr. K. (Department of Geography, King's College, London)

Willatts, Dr. C. (Ministry of Housing and Local Government)

Government Agencies

East Anglia Economic Planning Board (Mr. R. Mabey and Mr. C. Williamson)
Ministry of Housing and Local Government
Board of Inland Revenue
Board of Trade
Central Statistics Office
Department of Employment and Productivity
Department of Health and Social Security
General Register Office
Home Office
Ministry of Agriculture, Fisheries and Food
Ministry of Public Building and Works
Ministry of Technology
Ministry of Transport

Institute of Geological Sciences

The maps on pages 68 and 69 are based on data supplied by the Institute of Geological Sciences and are printed by kind permission of the Director.

Local Authorities

Birmingham City Council, Planning Department
Cambridgeshire County Council, Planning Department
East Anglia Consultative Committee
East Sussex County Council, Planning Department
Greater London Council, Research and Intelligence Unit
Hertfordshire County Council, Planning Department
Liverpool City Council, Planning Department
London Borough of Brent, Planning Department
London Borough of Southwark, Planning Department
Norfolk County Council, Planning Department
Norwich City Council, Planning Department
West Riding County Council, Planning Department

Others

British Rail
Centre for Environmental Studies
Coleman, Dr. Alice, King's College London
Eddison, T. and Wedgwood-Oppenheim, E., Institute of Local Government Studies, University of Birmingham
General Post Office
Hall, Prof. P. and Cripps, E., University of Reading
London Boroughs Management Service Unit
National Institute of Economic and Social Research
Nature Conservancy
Potter, Prof. A., University of Essex
TPI Computer Group
Water Resources Board
West Midland Regional Study

Ordnance Survey

References

Adams, J. M. 1969 "Mapping with a third dimension"; Geogr. Mag. 42 (1), pp 45 - 49

Bertin, J. 1967 "Semiologie Graphique"

Bickmore, D. P. 1968 "Maps for the computer age"; Geogr. Mag. 41 (3), pp 221 - 227

Bickmore, D. P. and Shaw, M. A. 1963 "The Atlas of Britain"; Clarendon Press, Oxford

Coppock, J. T. 1969 "An agricultural atlas of Scotland"; Cartographic J. 6 (1), pp 36 - 46

Gaits, G. M. 1969 "Thematic mapping by computer"; Cartographic J. 6 (1), pp 50 - 68

Hawkes, J. G., et al. 1968 "Computer mapping of species distribution in a county flora"; Watsonia 6 (6), pp 350 - 364

Lang, T. 1969 "Rules for robot draughtsmen"; Geogr. Mag. 42 (1), pp 50 - 51

O'Leary, M. et al. 1966 "Fortran IV and map program for computation and plotting of trend surfaces for degrees 1 through 6"; Computer Contribution 3, Kansas State Geological Survey

Perring, F. and Walters, M. 1962 "Atlas of the British Flora"

Robertson, J. C. 1967 "The Symap program for computer mapping"; Cartographic J. 4 (2), pp 108 - 113

Rosing, K. 1969 "Computer graphics"; Area 1, pp 2 - 7

Scher, M. 1969 "Orthophotomaps for urban areas"; Surveying and Mapping 29 (3), pp 413 - 422

Stubbs, G. M. 1967 "Point data and its uses". Regional Studies Association 3rd Conference. First Session: A co-ordinate referencing system for planners, pp 10 - 24

The following Government changes were announced in the press on 16th October, 1970. The Ministry of Technology and the Board of Trade are to be joined into a Department of Trade and Industry. A new Department of the Environment will incorporate the responsibilities of the former Ministry of Housing and Local Government, Ministry of Transport and Ministry of Public Building and Works. The present Department of Employment and Productivity will in future be known as the Department of Employment.

The announcement came too late to alter all the entries in Appendix 1; the changes will affect about 20 of the items covered.

Abbreviations

ADP	Automatic Data Processing
BCD	Binary coded decimal
BSO	Business Statistics Office
CB	County Borough: City Borough
CGIS	Canada Geographic Information Service
CRT	Cathode ray tube
DEC	Digital Equipment Corporation
DEP	Department of Employment and Productivity
DHSS	Department of Health and Social Security
EACC	East Anglia Consultative Committee
ECU	Experimental Cartography Unit
ED	Enumeration District
EPR	Economic Planning Region
FNR	Forest Nature Reserve
GB	Great Britain
GCD	General Claims District
GRO	General Register Office
HMSO	Her Majesty's Stationery Office
IBM	International Business Machines
ICL	International Computers Ltd.
IDC	Industrial Development Certificate
IGS	Institute of Geological Sciences
IMTA	Institute of Municipal Treasurers and Accountants
LA	Local authority
LNR	Local Nature Reserve
MAFF	Ministry of Agriculture, Fisheries and Food
MB	Municipal Borough
MHLG	Ministry of Housing and Local Government
MLH	Minimum List Heading
MoT	Ministry of Transport
MSS	Manuscripts
NAAS	National Agricultural Advisory Service
NCC	National Computing Centre
NFS	National Food Survey
NNR	National Nature Reserve
ODP	Office Development Permit
OS	Ordnance Survey
PATRAC	Planning and Transport Research and Computation Co. Ltd.
PAYE	Pay as you earn
RD	Rural District
RDC	Rural District Council
RGS	Royal Geographical Society
RHB	Regional Hospital Board
ROSPA	Royal Society for the Prevention of Accidents
RRL	Road Research Laboratory
SDD	Scottish Development Department
SIC	Standard Industrial Classification
UD	Urban District
UPCS	Universal Printing Counter Sorters

Contents

1

Section 1

Introduction: objectives and conclusions of the study

Section 1
Introduction: objectives and conclusions of the study

The first necessity for this report seems to be some brief and simple explanation of the processes of automatic or computerised cartography.

It is convenient to divide the subject into the headings of Data capture; Computer processing; and Graphic output.

Most data for mapping exist in graphic form – as field sheets, air photographs etc. To make them amenable to computer processing they have to be translated into strings of digits – i.e. numbers or grid references. This stage of digitising involves distinguishing those things which are lines or areas from those which are points: the latter can be located by one pair of x and y co-ordinates and the former by a stream of such co-ordinates at very close intervals (say 3 to a millimetre). The digitising of lines and points is quite a quick process on modern equipment – about twice as quick as "fair" drawing. But, not surprisingly, a great deal of data is produced, and magnetic tape is at present much the most economical medium for its storage. In order to distinguish on the tape between the line of, say, a boundary and a road, each stream of co-ordinates is proceded by a code number which the computer can recognise.

Computers are relatively unbothered by the quantity of data fed into them, and are able to present this very quickly, either in its full original detail or sorted or aggregated as instructed. Thus it is generally sensible for the computer to hold the data in all the detail that is ever likely to be required.

And because it holds only the basic data, it reaps the advantage of having only one set of records to update. But besides being good at "data processing", the computer serves another function in automatic cartography. When a map is being drawn automatically, the computer tells the drawing machine where to move so as to interpret the positions that it holds on magnetic tape. Now it is obviously quicker to draw with accuracy straight lines than twisting ones; and in fact the drawing machine is rather like a motor car being driven at night – the driver has to brake for a sharp corner. By looking ahead on the magnetic tape, the computer can foresee corners in the lines to be drawn and can regulate the speed etc. accordingly. At the same time the computer can select the kind of symbol – square or circle or whatever – that is to be plotted, or the thickness or dottedness of the line. So an important distinction between automatic and traditional cartography is that the computer holds information about position in a "pure" digital form, uncluttered by graphics, which can be selected, by the push of a button, to suit the scale and purpose of the map. And if maps can be made quickly, there are likely to be many more problems that can be usefully examined in cartographic form. Some of these will be ephemeral and will naturally demand different methods of display to those traditionally used for mapping definitive subjects – so a new flexibility seems to emerge without any need to sacrifice graphic elegance.

These points are, of course, discussed and illustrated in the body of this report. But it was a distant glimpse of the possibilities of applying these ideas and techniques to the field of planning that gave rise to the project.

Terms of reference

The terms of reference for this study were laid down by the East Anglia Regional Office of the Department of Economic Affairs as follows:

An examination of new techniques in computer controlled cartographic and graphic display, and an assessment of the extent to which available, or existing, equipment and techniques could be applied to the problem of improving methods of presenting planning data. A comparison will be made of the relative costs in equipment and staff of the automated and traditional systems, together with the likely scope of benefits from the adoption of the former.

The investigation will be based on environmental information (economic and physical) available for one part of the East Anglia Region.

At the beginning of the project some fuller notes about the intentions of the scheme were set out, and it seems useful to quote these here.

(i) The main objective of the project is to examine developments in computer cartography to see whether data could be presented more quickly and more meaningfully than hitherto; to make cost comparisons, and inrelation to this work, to examine whether a computerised library of mappable information is feasible for East Anglia. The above investigation would be based on a pilot study of the Norwich area.

(ii) To fulfil these objectives it would seem to be necessary to have discussions and examination in the report of the following topics.

a An examination and assessment of new techniques and equipment

An examination of new techniques in computer controlled cartographic and graphic display, and an assessment of the extent to which available, or existing, equipment and techniques could be applied to the problem of improving methods of presenting planning data. A comparison will be made of the relative costs in equipment and staff of the automatic and traditional systems, together with the likely scope of benefits from the adoption of the former.

It might be particularly useful to show how better presentation techniques would improve policy decisions and speed up research work. Under this head some actual output using the Norwich area – perhaps a worked example of a sub-regional planning problem would enable cost comparisons to be made, and would demonstrate the type and quality of the output.

b Examination in some technical detail of the feasibility, methodology and cost of getting data into required format

A catalogue of information should be prepared, e.g. population censuses, Ministry of Labour data, censuses of production and distribution (within "confidentiality" limits), land use surveys, agricultural returns, traffic flows, river flow statistics, precipitation, etc. The reliability and the "resolution" of each type of information would be noted along with its frequency of revision and its present form e.g. field sheets, manuscript lists or punched cards. Location of the original documents and accessibility to them would also be noted. The report would also be concerned with the conversion of this information to computer compatible form, and involves three processes. First, the locational one digisting boundaries e.g. of parishes or even of built-up areas within the parish, in terms of which the information has been collected, and, secondly, of recording the quantities involved for each geographical unit. Thirdly, of computer processing e.g. up-dating of the information, comparing the merging of one piece of information with another; generalising the information for changes of scale or variations of reliability. The end product of this stage would be magnetic tapes ready for plotting.

There should also be discussion of aspects which would help to assess the feasibility of assembling a regional data bank. For example:

Can enough useful data be made available for those who would wish to use it?

The problem of historical data – going back to, say, 1951. Feasibility of getting this into required format.

The problem of confidentiality.

The problem of the referencing of historical and new data. What sort of areal basis to be used?

Cost and method of updating data bank.

c Assessment of current research

Clearly the need or demand for future research will have to take into account what else is going on in this field. Some assessment would, therefore, be useful.

d Need for a regional/sub-regional data bank

Under this head it would be useful to set out the demand for such a data system, at what level it would be used and for what purpose. It would be important also to discuss how a regional data bank would fit in with the national data collection process.

From this note, then, it will be seen that a very wide-ranging study was called for: it is not surprising that some of these objectives have not been achieved within the 12 months concerned. At least half the problem has been to assemble additional staff, and to equip them with sufficient technical knowledge of Automatic Cartography on the one hand, and of the sets of official and other statistics available for mapping purposes on the other: this specialised knowledge takes time to acquire, and at least part of our time has been taken up by a learning process.

Conclusions

Some of the conclusions from this feasibility study are self-evident: thus it is demonstrably possible to translate planning data into computer form, and to play it back as maps and, more precisely, as overlays to topographic base maps. There was doubt about the Unit's capacity to do this a year ago, and some of these doubts are now resolved; the Sample Maps also show the time scales involved. A Catalogue of statistical data potentially available for mapping has materialised; and it will be clear that, if such a document is found to be useful for planners, then its regular maintenance should be a function of, e.g. the Central Statistical Office, rather than of a university research unit. Section 5, in describing some of the techniques of automatic cartography, implies the need to see that there are links between the different experimental approaches to this problem: and links, rather than any superimposed uniformity, at this stage.

That these points are obvious is not to denigrate them; however, there are some other conclusions too.

Geo-coding

The subject of geographic coding or referencing seems to be something of a new discovery for planners. But even if some data do materialise neatly packaged with their appropriate geographical references, it seems inevitable that there will be many other data that do not; there will, for example, be new parcels of land as a result of new road construction, etc. and the resulting new geographical shapes will have to be assimilated into the "system". This will involve digitising – and digitising lines as well as points.

Digitising has been one of the fundamental problems of automatic cartography for a decade, and some experience of this technology does now exist. It is important to ensure that this experience is available to planners. In so far as this itself leads on to the purchase of digitising equipment and the production of magnetic tapes, it also points to the need for the effective organisation of cartographic data banks. This is a large and technical subject, liable to involve substantial expenditure on equipment and research.

Data exchange rules

As any organisation advances in its interpretation of an information system, so it seems important that data tapes should be "exported" to be played back on the equipment of other organisations. This assumes acceptance of a common format for the exchange of the cartographic data. Possibly planners should be aware of the current discussion of this subject by the Royal Society's Cartography sub-committee. (See Appendix 2)

Legibility

On the output side of automatic cartography there is also a need for further research. In particular, it is important to establish how much of the information presented in map form has any prospect of being read and understood even by the best-intentioned user. There has been little scientific work so far on the evaluation of normal cartographic displays; but there has been even less on the effectiveness of presenting, e.g., rateable value, in the form of a contoured surface, or on the relative merits of topographic maps or orthophotographs as the base map. At at time when the quantity of planning maps is increasing greatly, there is a real need to study the degree to which the language of cartog, aphy is capable of being understood.

Too many maps?

The ease with which computer maps can be made also involves its own dangers. It is all too easy to produce from the computer a hundred maps and more, whereas by manual methods only one could have been produced in the same time. But the hundred maps may answer the particular question no better and be much more lengthy than a single map. One is back to the problem of a discussion with a knowledgeable but loquacious bore.

The economics of automatic cartography

This study has gone only a small way to evaluating the cost effectiveness of automatic cartography. While some hard and reliable figures on this subject are greatly to be desired, it is clear that far more thought needs to be given to the formulation of meaningful questions about costs in relation to existing planning organisations, computer installations, etc.

Finally, let it be said that no apology is made that this scheme is in the field of cartography. The report is written from a cartographic point of view rather than from a planning one. There is very considerable use of maps in making planning decisions, both at the Local Authority and at the Regional Planning levels – and this, in spite of the slowness and expensiveness of existing cartographic methods, and the fact (emphasised later in this report) that most Central Government statistics are collected for presentation as statistics and not as maps. We do not attempt any argument about the merits of one or other method of communicating information; both maps and statistics should be equally available and available with equal rapidity.

Section 2

Automatic cartography and data

Section 2
Automatic cartography and data

Speed, resolution and cost

There exist a number of important British maps and atlases in the field of regional planning. Pride of place in this context should rightly be given to the Atlas of East Anglia published by the East Anglian Consultative Committee, 1968. Mention must also be made of the distinguished series of Planning Maps (at 10 miles to the inch scale) and the smaller scale "Desk Atlas" prepared by the Ministry of Housing and Local Government. Mention too, should be made of the "Atlas of Britain" (Clarendon Press 1963) and of the "Atlas of London" (Pergamon 1968). And outside Britain there has been a wealth of atlases for regional planning produced in the last 10 or 20 years – The Atlas of the Resources of British Columbia, L'Atlas de la France de l'Est, the Urban Atlas of 20 American Cities, the atlases of the Ruhr, of the Ukraine, of Czechoslovakia, of Hungary, of Japan, of India. The list is a long and distinguished one and the net cost over the last 20 years must run into many hundreds of millions of dollars.

All of these productions have the obvious merit of presenting the resources of a region in a synoptic manner which makes an immediate appeal to human peripheral vision – a quality that seems to have little application in scanning columns of figures. But there is an equally obvious short-coming in the out-of-datedness of maps. The production time for the Atlas of Britain was 10 years, and it is common practice that single maps need anything between six months and two years, to translate from raw data to printed form. A "response time" of this order of magnitude must condemn cartography to handling only those problems which are sufficiently non-immediate to be able to afford such relaxed treatment. Decisions on many planning matters simply cannot accept a time scale which assumes making decisions on information which is so much out of date. Raw data for planning requires to be translated into map form in a period of between one day and one month. So speeding up by a factor of 10 - 100 needs to be introduced into cartographic methods if they are to keep pace with planning needs: this seems to be the first new commandment.

In any discussion of the speed of cartography, note must also be taken of the quality of map that is involved. We do not necessarily argue for a rigid adherence to graphic specifications that have been used in planning cartography over the last 20 years – though the merits of retaining a reasonably well understood symbolism and hence preserving the existing language of cartography, are not to be dismissed lightly. We do, however, draw attention to the need for high resolution cartography, capable of handling minutiae of detail on the same kind of basis as on Ordnance Survey maps. It is accepted that the human eye can interpret this scale of detail; and we have aimed our form of Automatic Cartography to match this kind of topographic detail.

Whether or not base maps are produced automatically, it is clear to us that the information to be overlaid on them should be comparable in scale of detail. This means that Automatic Cartography has to go beyond placing symbols in the generalised area (whether grid square or centroid) of houses, parishes, factories etc.; it must be capable of pin-pointing and of delineating with a precision equal to that of the base map. The process of doing this may take fractionally longer than the use of more approximate methods such as SYMAP, and it is not suggested that what has been called precise Automatic Cartography is a substitute for techniques like SYMAP; nor indeed vice-versa.

This study may suggest that the geographical resolution and the speed which automatic cartography can now attain are in fact likely to meet many planning needs. The map section and the times shown to prepare the sample maps must speak for themselves.

Another factor is cost. We were asked to assess the comparative costs of the old and new cartography. But often the two are not really comparable, and anything other than the most superficial comparison is quite misleading. With this qualification, however, it may be worth noting that ten years ago the production of the Atlas of Britain cost approximately £150,000. It involved 200 pages of maps each 20" x 15" i.e. 50/- per square inch. The preparation of the present map section might perhaps be said to have cost half of the total research fee – i.e. £2,500 – and to have produced some 1400 square inches of "map" i.e. 36/- per square inch. From such over-simplified figures it is possible to suggest that the cost of preparing these sample maps has been about two-thirds of the cost of those in the Atlas of Britain. Having said this, however, the qualifications mount formidably, whether it be, for example, in terms of allowances for capital equipment, or in comparisons between production in a plant and experiment in a laboratory, or in inflation over a decade.

At least there is nothing to indicate that the cost of automatic cartography is higher than that of traditional cartography and much to emphasise that it is far quicker.

Although it does not seem to us to have very much significance on the economics of automatic cartography for a regional planning agency it should, at this point, be added that the capital cost of the equipment in use by the Unit is of the order of £100,000. Let us from this point of departure speculate further, bearing in mind, of course, that decisions about general purpose computer purchases – regionally or centrally – are likely to precede and influence subsequent decisions about systems of automatic cartography. If this equipment were to be bought at today's prices and specifically with efficiency in a production, rather than a research role in view, these costs might increase by some 20%. If spread over a 6 or 7 year period they might be budgeted at some £20 k.p.a. To secure maximum output, round the clock, a staff of about 12 might be economic – say a further £20 k p.a., plus some £10 k p.a. for materials and colour proving, map printing, etc.

We have estimated in another context that this notional budget of £50 k per year might be capable of an output of some 2,000 different maps (on the basis, for example, of land use maps; perhaps in quantities of several hundred of each map; and in a size of say 20" x 15"). Against this figure of £25 per map the qualifications and assumptions again pile up. The availability of some form of base map on which to overprint this information would be assumed; the 2,000 maps a year would also assume a range of scales for a particular area (say from 25" to the mile to 1" to the mile); some maps in many colours, more in few. The availability of the raw data on land use or rateable value or whatever topic is also assumed and immediately raises the questions of how long it would initially take to get the data into computer readable form and thereafter, how long it would take to keep it updated. There are some clues to this first question from our work for the sample maps.

The digitising of the Land Use maps (i.e. the boundaries of parcels with coded descriptions of what was either side of each boundary) took 26½ man-hours for these 4 sq. kms. of suburban and 2 sq. kms. of rural land. Given weights of 8 for urban, 4 for sub-urban, and 2 for rural areas, because of the different densities of detail, it is estimated that the 12,420 sq. kms. of the East Anglian region would take 1300 times as long. The time required depends also upon the scales of map digitised, and extrapolation from such a small sample is somewhat speculative. It is suggested, however, that the digitising of the whole region might require of the order of 35,000 man-hours. In a production situation as suggested above there might be 3 digitising tables with pairs of operators working in 2 shifts: under these circumstances we guess that the digitising of East Anglian Land Use might take some two years. Since it took 35 man-hours to digitise the 3,786 hereditaments in the sample area, we estimate that it might take 6,200 man-hours to digitise those in East Anglia, i.e. about 4 months. Let us therefore allow three years for the digitising process for these two data elements for East Anglia. Digitising is the longest of the stages of automatic cartography and play back from one drawing machine should be able to keep pace with three or more digitising tables. Much of the information would, of course, be required at several scales, and in various different graphic forms. The prospect of output at anything like a rate of 2,000 maps a year is overwhelming to say the least; and it is of course obvious and proper that many of the maps would be ephemeral.

The question of cost seems to be one that can more fruitfully be looked at from other directions. First of all, cartography must be seen as one of several methods of information processing on which both planning and administration depend. These include the stages of data capture and of data processing and, cartography apart, there are good reasons to doubt the cost effectiveness of traditional methods of handling either of these stages manually with an army of enumerators and clerks. If there is a case for introducing a computer into the system anyhow (e.g. to handle Rates), then such a computer can extend its role into further stages of information processing. And if automatic cartography is one of the desirable stages of information processing for planning, then some extensions to the computer — and relatively minor ones in terms of cost — can serve this end with notable economy. Secondly, much of the information that is involved originates from central government. If this information were, in some utopian future, to be directly accessible to regional computers and coded in such a way as to make automatic mapping possible, then the costs of automatic cartography can be reduced still further.

But there seem other advantages to automatic cartography than those of speed and cost: these can be broadly grouped together under the heading of flexibility. Traditionally, it is assumed that comparisons between maps of different subjects at the same scale are conducted visually and memorised — attempts to use tracing paper tend not to be especially easy — but the correlation of detailed patterns is a significant concern to planners. Thus, there is a real need to overlay one distribution pattern on a variety of bases at different scales, some of them even being air photographs; others perhaps not being "base maps" in the traditional topographic sense but sociological patterns. The flexibility of a system of cartography in which change of scale or variations in graphics can be obtained at the press of a botton seems to be a significant factor. In rather the same way, where a complex map has been produced as a result of combining many variables, there is often a need to "unscramble" the data, and present the variables one by one or in some different combination: such matters are relatively quick and easy given automatic cartography. There is a third desideratum: it is the ideal of a "conversational mode" of cartography where the user asks a question of the computer which draws out the answer immediately. This answer may be on a cathode ray tube or, less spectacularly, though with greater precision, on a plotting table. Such a conversation may also demand statements of area or of distance; these also lie within the capacity of automatic cartography.

These notes defend the value of maps and in particular "instant maps" as tools in planning research and decision making. Our experiments can most usefully be seen as early, provisional steps in this field. It is difficult to give a quantified answer to "how valuable" without making the question much more specific. And they are useful even if they only show that the technology of precise automatic cartography has now reached a speed, a geographic resolution, and a graphic quality that should be taken seriously, and that might be economic.

Cartographic data banks

Automatic cartography can be subdivided into convenient subsets of technological activity — the Experimental Cartography Unit divides its research activities for convenience under the headings of data capture, computer processing, and output. But these subdivisions mask a common concern with the handling of data and these separate activities of automatic cartography acquire focus in terms of data banks.

It is of the essence that cartographic data banks will contain extremely large volumes of information. A recent study for a rectangular area of 10,000 sq. kms. (in South West Britain) concluded that 800 million characters of information — over 100 reels of magnetic tape — would be involved to record the reasonable basic essentials of topographical, geological, soil, ecological and planning data. These would be recorded and digitised where possible at a scale of 6" to the mile. The very tentative nature of this estimate of 1 reel of magnetic tape per 100 sq. kms. must be emphasised; it is an overall estimate and may grossly underestimate the situation in major built-up areas: nevertheless, it does provide one set of figures to work from.

Clearly magnetic tape is the only economic and first line storage medium for this kind of work in sight — however it suffers from the drawback that it permits only serial access. In the case of cartography it is certain that many parts of files will have to be retrieved more than once during an operation (e.g. a river being treated in its own right and as an administrative boundary). And the whole retrieval operation might be required within an hour, a day, or a week. Such an operation might require access to a significant part but not the whole of the 800 million characters involved for the 10,000 sq. kms. Under these circumstances the use of a large direct access storage device such as the IBM Data Cell (capacity 400 million characters) seems necessary. The relevant magnetic tape files would then be copied on to the Data Cell and, with a sophisticated indexing system, fresh tapes to answer specific queries could be produced ad hoc and very rapidly — i.e. without having to space past great masses of unwanted material.

It is at this stage worth emphasising the elementary but vital distinction between a data bank and a simple shelf-full of magnetic tapes. The distinction has some analogy to a library of books and periodicals, on the one hand with, and on the other hand without, a card index reference system. But the unorganised shelf of magnetic tapes is in worse state than the equivalent collection of books, because the tapes can only be machine read. Furthermore, the whole object of automatic cartography is to record each piece of geographical information once only. Thus the system of cross references must be elaborate, dependable and — in the nature of things — quick.

Assuming that the data banking system does offer this, what are the benefits that follow? In the first place revision can be handled on a very rapid basis depending on the fact that for each line or area or point there will be a unique set of records to up-date. But revision of any particular feature of a map is liable to have side-effects on other features. A new road, for example, may cross a river at a different point and set up new patterns of "parcel" boundaries. To carry out such revision will involve altering the files of a number of features other than roads. In the second place, there is an increasing interest in the comparative study of different parameters cartographically displayed: for example the angle of slope of a valley side in relation to the underlying geology or the probability of a particular amount of precipitation. The study of many such problems of great theoretical and practical interest (either impossible or at least tediously slow and expensive today) is made feasible with a properly organised cartographic data bank: in considering

these points it must be re-emphasised that experiments for handling large quantities of data demand the existence of data in these large quantities, and also in an appropriate digital form on magnetic tapes.
This argument leads to some fairly expensive conclusions, such as a further computer investment of perhaps £250 k. Clearly not every Regional Planning Board – let alone each Local Authority or Unitary Authority – would be able to invest at this level – and even if they could there would be a quite formidable need to impose compatibility of file structures etc. from some central position of eminence. Yet, at the same time, it is to conclusions of these kinds that the first steps in automatic cartography seem to lead. But perhaps there might be some kind of hierarchy within a national information system for planning. Thus, at the simplest level, the planner should have equipment so that he can digitise new information to feed into the system; but at the same time he should also be able to call up from regional data bank, tapes of particular topics from which he can himself derive his own maps – albeit simple ones – ad hoc. At the same time there should be another level – probably regional, possibly national – at which the regional data bank is located along with special purpose high accuracy cartographic equipment. The two levels of sophistication in automatic cartography may well be separated from each other by a cost factor of ten, but it is essential – and feasible – that conversion between them should exist.

The Experimental Cartography Unit having assembled its basic equipment for relatively simple mapping now contemplates some four or five years' work on problems of cartographic data banks. To some degree this formidable data handling problem may discourage planners from leaping too sanguinely to optimistic conclusions about the immediate blessings of automatic cartography. For example the first stage in such a process must be to digitise the positions of lines, areas and points, and this digitising has to be preceded on the tape record with codes or markers that can fit eventually into a comprehensive data bank file system. So perfectionism would suggest delaying any start on automatic cartography until the data management problem has been more thoroughly investigated, and even longer perhaps – until government decisions have been made about the whole administration of data. Such a timid approach might cut more ice if the need for better information for planning a whole host of location specific problems, from housing and motorways to the preservation of good landscape and the halting of pollution, could wait. But it cannot, and in the meantime therefore a series of wastefully imperfect data banks will come into existence. Under these circumstances perhaps the first plea is for the proper publishing of detailed work in this technical field, and for the acceptance of a standardised format for cartographic data on magnetic tape so that it can readily be exchanged between computer systems.

One final and sobering thought on this subject is that if cartographic data banking is unlikely to get itself organised on a national basis in Britain until the 1980's, we may by then need to extend the system into Europe as well.

Data for Planning

The Catalogue of UK statistical data potentially available for mapping – the Appendix 1 to this report – really explains itself. Nevertheless, it does mean necessary to draw attention here to a frustrating gap between the very many sets of statistics which the individual house owner or factory manager is called upon to supply, and the resulting data available to planners when it comes to studying patterns in anything other than the most generalised detail. There is, that is to say, a substantial loss of detail – and hence of information – in the pipe line.

There seem two partial explanations for this loss of detail; the first is the assumption – as it were from "above" – that generalised information is all that is required. The second is that the suppression of detail – and particularly of geographical detail – is in the best interests of those who supplied the information originally. To some degree it can be argued that the computer revolution has overtaken this situation. One of the merits of the computer is its ability to process very large quantities of original data with great speed and flexibility. If the original data are held in all their present detail, the computer should be able rapidly to present both the generalised summary and the detailed pattern for a selected area, in either statistical or cartographic form on request. This is something new; and in consequence the need for printing summary tables as the only documents which it is practicable to publish – the only visible tip of the iceberg – no longer applies: it is better to supply raw tapes. The computer should be able to apply screening programs to an area so as to pinpoint in the first place where there is a potential breach of confidential information, and then find a way round the problem, e.g. by some aggregation either of topics or of areas. The problems of confidentiality have been argued further in Section 3 of the report under the heading of Geographic Referencing. The points to be emphasised are:

a That much of the information that goes into official statistics is collected on a detailed, as it were, house-to-house basis, and for a good many purposes its subsequent aggregated form is far from ideal.

b That with the aid of computer screening many of these data, at present imprisoned under confidentiality regulations, could usefully be released.

c That the presentation of the detailed information by map will sometimes be more and sometimes less revealing than by statistics.

The question is not so much whether access to the details of the raw data can be achieved under adequate protection for the individual, but whether individual Government offices that have collected these data are prepared to regard such procedures as economically or politically worthwhile. This itself depends partly on the need for and value of, such data; and partly on the validity of the assumption that "privacy" is "good" and any infringement of it is "bad" (see Jacob: "Seven dangers of computers": New Society 11 Dec. 1969).

One of the sets of data for the whole UK, which is available in its full detail, is Rateable Value and it is for this reason that the majority of the sample maps have been based on Rateable Value data. The maps have concerned themselves both with displaying the data in full geographical detail (but generalised by class categories), and with displaying it generalised into, e.g. grid squares, postal code areas, etc. Unhappily, there are very few other Central Government data available in this unaggregated form, and it is surprising that more cartographic use has not been made of Rateable Value data. One of the reasons why Rateable Value data are available in detail is that rate demands have to be sent out by the Local Authorities to individual rate payers. It is interesting to find that the City of Norwich has its Rateable Values in computer form, arranged in sequence of streets. The particular arrangement that has been adopted in this case is doubtless convenient for the immediate purposes for which it was designed — i.e. payment of rates. With some adjustments, such a file could also usefully be the direct basis of mapping. The main need here would be to record Rateable Values of hereditaments in the same sequence as that in which the geographical positions of the equivalent hereditaments are digitised: the merging of values with positions seems to be a simple subsequent operation.

Furthermore, it might well be convenient to arrange the sequence of the two computer files (values and positions) into the groupings of the new Postal Codes. This would then provide one convenient method of aggregating them, and in addition, would seem to suggest simplifications in the postal problems of sending out rate demands, etc. A Postal Code Unit is the smallest part of an area that can be identified for machine sorting. It contains not more than one street, or one hundred letters in a section of that street, or not more than seventy possible calls in any section of a street. Addresses receiving more than twenty items get an individual code in any case. Postal Code Units seem to form a convenient basis for aggregation of various other sociological and business statistics; this seems especially so where it may be possible to arrange collection of information by post and on pre-coded forms, without the use of door-to-door enumerators. (It is on a postal basis of this kind that the US Bureau of Census collects its population information.) Even allowing for sorting out individual mistakes, the saving by not using enumerators would seem substantial. It should be added that aggregation of addresses into postal codes in no way should preclude aggregation into other patterns where such aggregation is necessary, e.g. into grid squares, or into three-dimensional surfaces displayed by contours etc. Here again the computer should make it possible not to have to opt for only one of the several alternatives; it should give flexibility to information processing and at a high level of resolution.

Automatic cartography will, within the next two or three years, be able to present dynamic in addition to static maps – maps that can actually show changes over a period of time by computer generated films, using animation techniques. By such methods it will be possible to extrapolate trends into the future on a detailed regional basis. However there are well known geographical problems in relating one census to another. Any geographically detailed work on trends is bedevilled by the changes that take place between the boundaries of different administrative areas and thus confound the whole issue of comparison between dates. If, in a utopian future, each piece of information is not only aggregated into administrative areas but is also coded with a unique geographical reference, all the very obvious advantages of changing boundaries can be reaped, and at the same time, the basic information can be re-aggregated to fit the new boundaries for those rather few instances where aggregation or generalisation is really necessary. Under such circumstances the question of whether aggregation is carried out to grid squares, administrative boundaries, postmen's walks or to completely arbitrary shapes (e.g. the likely route of a new motorway) makes no difference; the planner can choose what he wants when he wants it. Such flexibility immediately has its possible advantages in connection with the confidentiality problem, but it is only feasible given a system of basic geographical locations.

The question then arises as to what is "basic geographical location". In Section 3 attention has been drawn to the fact that at one scale a well, for example, may be treated as a point, whereas at a somewhat larger scale it is seen to be an "area". Arbitrary decisions have to be made about scale; e.g. the Ordnance Survey's decision to map urban areas at a scale of 50" to the mile. Here is one agreed level of detail. Planning information is concerned both with plani-metric shapes – of buildings, parcels of land, streets, rivers, etc. – and with statistical quantities associated with those shapes. The ideal in geographic location would seem to be that the shapes should be shown – as they mostly are – on the basic national survey maps, and that these geometric shapes should be held in digital form in a computer based data bank. Associated with each shape there should also be a code number which not only indicates its "species" but also serves as a unique reference. This code can then serve as a link with any appropriate statistical quantities.

Whether or not this cartographically ideal situation will have materialised by the turn of the century, it clearly will not have done so – at least for large areas of this country – during the 1970's. So some compromise is necessary. It may, in fact, be sufficient to choose a point of location for a house or a small hereditament which is roughly in the centre of the shape. On this basis – as the Sample Maps indicate – a symbol can automatically be drawn and overlaid on the base map, so as to identify unambiguously the topographical shape referred to. Point locations of this kind will certainly provide a geographical keying system for a large number of the small features to which the original data refer, but they are much less satisfactory for larger areas or for linear features such as roads or rivers. And, even in combination with a large scale base map, point location may be of little help in defining property boundaries. However, it is technically no more difficult to arrange for the location key to be a segment of a line (or a series of segments representing the outline of an area), rather than a point. Any regional system of geographical coding that was based on points only and not on points, lines and areas, would be inviting trouble; and needlessly, since lines and areas can be digitised just as accurately and quickly as points.

The Catalogue shows that a number of data sets regularly collected by Government departments is already in computer form. As with so many computer problems, however, computer compatibility is more an intent than a fact. Perhaps gradually Government statistics for planning will be centrally organised so as to impose a greater compatibility between the computer formats of various data banks. An alternative, which may well be less unattractive, is the development of translation programs which will allow both reasonable freedom to particular installations, and the possibility of transferring information to quite seperate computer systems.

Section 3

Geographic referencing

Section 3
Geographic referencing

Referencing possibilities

Before any maps can be produced, some sort of geographic referencing must be attached to data. So it is necessary to consider the relative merits of different possibilities for geographic referencing (or 'location coding') with particular reference to planning maps. The question of geographic resolution, which greatly affects the meaningfulness of the maps which can be produced must also be considered.

We start from the simple topological point that features on the earth's surface may be referenced in terms of zero, one, or two dimensions, i.e. to points, to lines, or to areas. (Volumes may be relevant to some planning work, but it seems appropriate to leave them aside in this report.) Though each possibility may be appropriate for certain features, there is no clear division in practice since the appropriate dimension is affected by the scale of consideration. For example, a well is a point on a topographic map, but an area on the ground: a railway is treated as a line at small scales, but it too is an area. A block of houses may be considered as area, line or point, again depending on scale. Sometimes it is necessary to represent areas as points because their outlines cannot easily be established, e.g. individual farms may sometimes have to be represented by farmhouse centroids.

For computer handling, all of these geographic references have to be expressed in terms of co-ordinates. Points are represented by single couplets of co-ordinates; lines, by chains of couplets; and areas, by closed lines or by chains of lines, each of which is in turn defined by chains of co-ordinate couplets. There has been a tendency to regard points as easier than lines or areas to reference in these terms. Certainly when co-ordinates are to be read manually off a map, punched into machine-readable form, and checked, it is much less work to deal with the single couplets associated with points. Also, in primitive card-oriented computer systems it was easier to handle a fixed number of points on every card than to handle lines made up of variable numbers of point couplets, probably not filling the final card.

These restrictions, however, have been removed by the development of efficient analogue-to-digital converters (digitising tables), and of modern computing systems based on the use of magnetic tape. The d-mac digitising table permits points, lines, or area outlines to be digitised without manual co-ordinate reading and punching, and with no less accuracy. Points are digitised by locating a pencil in the correct position and pressing a switch; lines and outlines, by holding the switch down while they are traced out with a pencil, and concluding by pressing an "end of line" button. Given the fact that everything (whether point, line or area) has to be labelled, by typing one or more codes, and that these codes must be checked, the total effort in recording a short line or outline is little more than that for a point. The effect of improved technology has been relatively much greater on the recording of lines and outlines than on that of points, and it is necessary to realise fully the flexibility which is now possible.

Before computing became important, features were most commonly referenced to irregular areas, such as parishes or counties, or to lines such as roads or railways. The use of areas fitted in with a classifactory mentality, and with the view then held of geography as areal differentiation. Science has now moved towards the use of quantitatve scaling in preference to classification, and geography deals chiefly with spatial variation and interaction. The implication of these changes is to favour the finest possible geographic referencing, coding data in its original form rather than aggregating it in codes for larger units.

The Flora North America project, for example, records the geographical co-ordinates of each point observation, rather than codes for states or natural regions. The reasoning is that enquirers are likely to be interested in areas of various shapes, which can be specified as polygons for the purpose of searching the point codes. Use of predetermined areas is much less flexible, and presents the enquirer with a much greater problem of sorting the data he desires from that provided.

On the other hand, if detail is relevant, for example, for land use where areas may need to be calculated, it is not necessary to treat areas as points; there may be little extra work involved in digitising the whole outline of a property, field etc. The present state of technology, then, does not impose great constraints on the type of geographic referencing to be employed; the type used should be that most suited to the data and to the purpose envisaged. The only constraints are economic, and computational in that some programs or algorithms which might be used may be limited to handling one type of referencing.

Considering Britain alone, the National Grid seems to provide a very useful frame for geographic coding. Its units are of approximately equal area, and are now used to define the boundaries of OS maps. Since conversion from digitiser table co-ordinates is achieved by giving the exact co-ordinates of corner points when OS maps are digitised, the National Grid system is very convenient for automation.

The international standard, however, will probably be 'geographical co-ordinates' – degrees of latitude and longitude, decimally divided. These are necessarily involved in projection changes. Hence, when data are to form part of a bank of international scope, they will probably be converted to geographical co-ordinates.

Almost all statistical information is collected on a detailed basis, for individual households, factories, offices or businesses. Unfortunately, it is not usually available in this form to users. Instead, it is aggregated to larger or smaller geographic areas or classes of phenomenon. The purposes of such aggregation seem to be: (i) to cut down the sheer bulk of numbers which must be processed and published; (ii) to reduce data to a comprehensible quantity by generalising; (iii) to make sample data meaningful by aggregating to classes which have sufficiently large samples that confident statements can be made about the classes; (iv) to conceal data that must be handled in confidence.

Purpose (i) is of much less importance when processing is done by computer. Purpose (ii) is less relevant to mapping, which in itself increases the comprehensibility of data. Mapping permits portrayal of complex patterns, which can be grouped mentally even when the individual occurrences cannot. Some generalisation is achieved through representing exact statistics by graphic symbols. Purpose (iii) is thoroughly necessary for sample data, even when they are to be mapped. Purpose (iv) is a complex issue which will be discussed on the following pages. This will be followed by a discussion of different methods of aggregation.

Confidentiality

Apart from the cost of coding, geographical resolution is restricted by the fact that official statistical data are often collected on the assurance that they are handled 'in confidence'. For many government surveys disclosure of information relating to individual units, such as households or businesses, is expressly forbidden under the terms of the enabling regulations. (An exception may be made where the individuals involved give permission for otherwise confidential information to be disclosed.) Confidentiality is safeguarded by aggregation on a subject or geographic basis or both. Where aggregation is based on som predetermined unit – for example, particular categories of the Standard Industrial Classification (in the Census of Production) or parishes (in the Agricultural Census – selective suppression or amalgamation takes place if there is only a small number of cases in a group. Different criteria are used to assess what constitutes a group that is so small that there is risk of disclosure. In the Agricultural Census consideration of the risk of disclosure is based simply on the number of cases, whereas in the Census of Production, the relative size of the business is also taken into account.

Where there is thought to be risk of disclosure, different procedures are followed. In the Census of Distribution all information relating to a cell with too few units is suppressed, whereas in the Census of Production only the number of cases involved is suppressed. In the Agricultural Census, summary data are available for parishes where there are three or more farms, irrespective of size. Where there are only one or two farms, the data are amalgamated with the data for an adjoining parish. The minimum sizes of grouping specified by the authorities to safeguard confidentiality are not in themselves unduly severe even from a cartographic point of view, provided that some flexibility is possible in grouping detailed subject classifications. For the four main censuses (Population, Distribution, Production, Agriculture) minimum cell sizes in the order of 3 to 6 basic units have been laid down, and cells as small as this should prove reasonable for a large proportion of mapping requirements. Achieving this degree of mapped detail at present is prevented more by problems of inadequate resolution in the basic geographic subdivisions in which the data are published, than by the actual rules governing breach of confidentiality. The Atlas of Britain (Clarendon Press 1963) contained a large series of maps at 1 : 2,000,000 scale dealing with the distribution of industry and based on statistics from the Census of Production : this raised confidentiality problems over which the Board of Trade were extremely helpful. The locations of factories were initially plotted within the Board of Trade in terms of employment and at 1 : 1,000,000; where there were less than three factories within the plotted location on the draft map, the Board of Trade asked the companies concerned whether they objected to inclusion. In the vast majority of cases (93%) assent was promptly forthcoming.

Confidentiality restrictions affect what is made available to the public, but vary in their application to government departments. Under the Statistics of Trade Act, an official of the Board of Trade would be guilty of an offence if Census of Distribution information relating to an individual unit were disclosed to a member of another government department, unless legal proceedings were involved. On the other hand, with suitable safeguards, confidential material from other sources, such as the population census or the employment returns, can be made available to selected bodies for selected purposes, as long as no subsequent disclosure occurs.

By contrast local authorities, industry, etc. generally have to make do with the published statistical summaries whose geographic subdivision is all too often quite inadequate, while national agencies, whose responsibilities may be much more general, have access to detailed confidential data.

In some ways, then, information detail available is inversely related to that needed.

To improve this situation, one approach is to attempt finer resolution of published data. There is a limit to the degree of geographical resolution that can be achieved without risk of disclosure. The territorial size of units satisfying confidentiality requirements varies from subject to subject, depending on the number and distribution of individual information units. Population information for a large block of flats would probably satisfy confidentiality requirements, while details of plant production in a specialised type of manufacturing could involve disclosure in any geographic analysis below national level.

There are many cases, however, where a more detailed level of geographic resolution for published material would not infringe disclosure rules. In the population census, the Census of Production, the Census of Distribution and local employment estimates, there is scope for further geographical subdivision of the cells for which information is published. There is only a small number of classes (e.g. small rural parishes, specialist types of manufacturing, some small local authorities) which could not be subdivided without disclosure. It would also be possible to trade off details of subject classifications against geographical detail, for example, by amalgamating SIC orders in the Census of Production to permit more detailed geographical subdivision. For household information, finer units such as post codes or 100 m. grid squares could be used in place of wards.

A second approach is through greater flexibility of data manipulation. Aggregated data cannot usefully be compared with data from other sources that have been aggregated to a different set of boundaries or categories, except in the cases where one set forms a sub-set of the other. It would, therefore, be a great advantage if the data could be re-aggregated in a variety of ways for comparison with other data, without necessarily lowering confidentiality thresholds. The key to such flexibility lies in retaining data in a basic unaggregated form, incorporating location and information codes so that the items could be sorted and re-sorted at will. This would permit re-aggregation for a revised classification, or to fit boundaries common to another variable, or for some more complex geographic analysis that safeguards confidentiality; for example, contouring, or measuring distributions such as trip lengths. Users would be able to specify the form of aggregation required, and would receive non-confidential summary totals and analyses from the operators. Where the machine-readable unaggregregated data were confidential, it would have to be guarded as securely as the questionnaires now are. Additionally, it would be necessary to safeguard against disclosure by deduction from overlapping aggregations. This would strongly limit the number of non-nested aggregations possible, and it might be useful to foresee all permissible small unit areas. New areas would have to differ from an aggregation of previously-used areas by more than one confidential unit, otherwise information about the latter could be derived by subtraction. An added safeguard would be to give individuals or organisations the option of receiving a "bank statement" print-out of (i) any information that had been divulged concerning them, with (ii) the reasons for the enquiry, and (iii) the body to whom the information had been given: this however might be very expensive.

Aggregation summaries would also be retained, so that a library of non-confidential summaries on a variety of bases would gradually be accumulated. Such summaries (especially those on a geographic basis) would also act as gazetteer giving totals, and serving a references to the basic (confidential) data which could be consulted for more complex analyses. Obviously non-aggregation is the most suitable approach when data are handled by computer.

A third possibility is to adopt a more open attitude to information. The statistical and cartographic difficulties cannot be overcome completely so long as confidentiality remains. Data which are confidential in one context are sometimes not so in another. For example, local authorities do not have access to Board of Trade figures for floor space; but they have the legal right to enter buildings and measure it for themselves. This is obviously ridiculous, and suggests that regulations covering confidentiality could often be relaxed.

Even if some statistics remain confidential except in aggregated form the more that can be freed, the easier will comparison become; aggregation can then be to the units used for the confidential information, and comparison will be difficult only between confidential information on different areal bases.

While it might be very useful if the statutes covering confidentiality could be repealed, this seems eminently unlikely in an age which is becoming increasingly suspicious of the computer. It should, however, be possible to make a more careful distinction between confidential and non-confidential information, even on the same questionnaire. There should perhaps be an enquiry into what data are regarded as confidential by the public and by businessmen.

Worthwhile improvements could be achieved immediately by pursuing the first approach as there is plenty of scope for finer geographic resolution without risk of disclosure. Many parishes, enumeration districts, and SIC categories could be sub-divided without revealing details of individual firms and families. The second approach would yield greater benefits still, as it largely avoids the problems of pre-digested summaries, changed boundaries, etc. It would permit comparisons in type, place and time. It also avoids a head-on clash with confidentiality as it is merely re-manipulation and re-aggregation of data; but it poses difficulties in maintaining confidentiality. In the long term, the third approach could be even more rewarding.

Unit areas for aggregation

There are two types of area which may be used as aggregation units (i.e. unit areas, from the point of view of the user of published data). Firstly, there are irregular areas defined for some other purpose, usually administration: the Data Catalogue (Appendix 1) shows that the most important type used is the parish – local authority system followed by the employment exchange system. Many other divisions are important at the regional level, which is not of prime concern here. The variation of such units in area and shape greatly restricts their value as unit areas for mapping, for geographic analysis or for spatial statistics, and to some extent even for conventional statistics; their deficiencies are not removed by contouring, since the resulting maps may be highly unreliable. There is no satisfactory way of comparing figures based on different irregular subdivision of area: the assumption of areal homogeneity implicit in disaggregation (to areas of overlap, or to a common system of grid squares) is almost never valid.

To take account of changes in human geography, administrative boundaries are frequently modified, at irregular time intervals; occasionally they are remodelled completely, as in Greater London. Hence comparisons over time are complicated, and it is necessary to keep stocks of special maps showing boundaries. In an automated system, the initial task of digitising these irregular outlines would be considerable, and there would have to be an efficient up-dating system. The advantages of aggregation to such areas are, first, that statistics for these particular areas are often of intrinsic interest to the authorities which administer them; and secondly that they are often made up from sub-sets of Real Estate boundaries which are of great legal and cadastral significance.

Alternatively, a set of squares may be used. In the British context, it is obviously convenient that these should form part of the National Grid (which is fortunately on a metric basis). Grid squares are already defined on Ordnance Survey maps; they are permanent; they can easily be aggregated to larger grid squares; and their equality of area permits direct spatial comparison of totals without calculation of densities or of percentages of area. Their equality and regularity permit figures for them to be manipulated as matrices, and used as the necessary starting point for many techniques of spatial statistics, including the mathematical modelling of diffusion. The regular spacing of control points is particularly useful for contouring, trend surface analysis, and spectral analysis. Grid squares provide a basis on which different distributions may be interrelated, whether they are expressed as small areas or points, or as larger areas of qualitative uniformity.

Maps can be produced automatically from values for grid squares, without any boundaries being digitised. Grey scale or point symbol maps can be produced on tabulators with a square print format, on cathode ray tube displays, or on flatbed plotters with projection of symbols, without any further transformation. On the other hand, single grid square figures are usually of no intrinsic interest, and are not simply related to real features on the ground such as streets or blocks; hence they are not suited to very local studies. They can be aggregated in order to approximate larger administrative units, but the error involved depends on the size of square in relation to the size of the units.

Since the grid squares are defined in terms of rectangular co-ordinates, they may be regarded as a coarse form of point coding, in which most possible co-ordinate pairs are occupied. Conversely, point coding may be regarded as the use of small grid squares, of side equal to the resolution of the coding; most of these squares are left empty.

As Stubbs (1967) has emphasised, the usefulness of grid squares as unit areas for aggregation and processing should not be confused with usefulness for coding of basic data units. If point co-ordinates are used for the latter, they can easily be aggregated to grid squares. On the other hand if irregular areas are to be used for aggregation, it is best to record their labels with each basic unit.

Data which can be treated as point data may alternatively be aggregated to lines. Lines can be relatively 'natural' features, meaningful on the ground and related to movement and delivery. They are easily digitised, and particularly suitable for output on a plotter; simple line maps can be produced cheaply and quickly. Streets can be used to aggregate household data, but since some streets are very long they must obviously be subdivided. The American Census uses a block face as its basic unit, i.e. a street frontage between intersections. By allotting addresses to block faces of which the termini are digitised, it has developed a powerful tool for mapping data which is otherwise almost unmappable because its only geographic reference is an address.

In Britain, we have the advantage of a detailed and well-advanced postal coding scheme devised for the machine sorting of mail (in the USA zip codes serve the same purpose, but cover larger areas). Codes are allotted to streets or parts of streets containing not more than 25 actual calls or 70 possible calls, in such a way as to make possible future insertion of codes for new developments, as well as revision of delivery arrangements without changing codes. A separate code is given to each delivery point receiving over 20 letters per day, and 20 to 25 codes make up a postman's walk (delivery round) in the average urban area. In rural areas, there are many cases where isolated households receive individual codes, regardless of their mail delivery; again, this provides flexibility.

The code consists of an 'outward' part of one or two letters followed by one or two digits, representing a sorting office, and an inward part of one digit and two letters, representing a street or part of a street. An outward code area has a capacity of 4,000 inward codes (the limitation is due to the sorting machine, not the number of characters involved), but will contain no more than 2,000 when first laid out.

For the aggregation of household and other highly localised data, postal codes have several advantages over streets. They are less geographically extensive, and tend to be more equal in importance – admittedly in terms of mail rather than population. More important, codes are already being allotted, and a file containing the postal code of each household is being written on magnetic tape. These codes will in future be necessary for questionnaires sent through the post, and so a form of location coding much more manageable than an address will be ready-made.

Postal codes, then, form small, compact geographic units for aggregation, related to real features on the ground. The task of digitising their locations would not be very great, since many of them may be represented as straight lines, and the task of inserting codes would be much easier than for individual households. The same code permits aggregation to postal sectors and 'machine towns', though these unfortunately are not related to any administrative boundary line. Postal codes are preferable to irregular areas such as enumeration districts, and have some advantages over grid squares.

A topic of obvious importance is the ease with which point, line and area data can be interrelated in digital form. This affects the desirability of digitsing data in the various forms; it is also a necessary consideration in that whatever form is most desirable, the system will have to cope with some data in other forms.

Points are easily related to grid squares by dividing the point co-ordinates by the known ratio between their resolution and that of the grid squares. In the simplest case this involves dropping the last digit or two; sometimes scaling or shift of origin may be necessary, but the process is very elementary so long as co-ordinates on the same grid have been used. Otherwise, there is the added complication of a projection change, but even this may not be onerous.

If, however, the points represent areas, they should be aggregated only to considerably larger areas, so that little of the area they represent falls outside the area to which they are allotted. Otherwise, the advantage of grid squares in being equal in area is lost, and the resulting statistics and maps are misleading. For example, the 1 km. squares used for LINMAP maps of SE England are too small relative to the parishes whose totals were allotted to centroids: the 'no information' squares are not areas without information, and they spoil the visual effect.

Points may also be related to lines, by testing whether their minimum distance from the lines is less than a given tolerance. The speed of this operation is inversely related to the number of points needed to define the lines which must be tested. It is more difficult to relate points to irregular areas, because of the large number of points needed to define irregular outlines. The problem is feasible for simple polygons, especially if they are convex.

Lines, too, can be related most easily to grid squares, by testing for when either x or y cross the grid intervals. This, however, is disaggregation, a statistically dubious procedure. If the grid squares are sufficiently large that only a small proportion of the lines cross their boundaries, this objection does not arise. Lines can similarly, but with much greater difficulty, be related to irregular areas. Lines can easily be simplified into points, by taking either mid-points, centres of gravity, or half-way between the terminal co-ordinates. It is difficult to relate irregular areas to anything else, but for the purpose of qualitative comparison they may be disaggregated, or simplified, into grid squares as in a line printer choropleth map. This can be achieved by first interpolating points where the outline of the area crosses grid lines (whose intersections are the centres of grid squares), and then arranging these points in scan order, so that those intersections which are internal to the area can be established. While such disaggregation may not be realistic for statistics, it is useful for qualitative data, such as outcrops of geologic formations.

In conclusion, the use of points is the form of referencing most easily related to other forms. The actual interrelating is most easily performed in grid squares.

Most maps of planning information are more meaningful when superimposed on a topographic base map, and so the automation of their production should not be considered in isolation. On the other hand, the automation of topographic mapping implies that availability of the new digital information may be relevant to the way in which planning information is recorded, and that it will at least be possible to select the topographic components which form the most relevant base for a particular planning map. On the other hand, the technique of automatic air photo rectification makes available a new type of topographic base, the orthophotomap. This has been found extremely relevant to the needs of planners and government agencies in Washington DC (Scher, 1969).

If the Ordnance Survey were to derive small scale maps by digitising selected features from large scale maps, it would at least provide planners with transport routes in digital form. If it were to go further and derive all maps from digitisation of field sheets, or from co-ordinates established during survey, this would provide much more information of relevance, such as house outlines. Hence it might not be necessary to digitise a point for each house. Experiments are now being conducted jointly by the ECU and the Ordnance Survey on the technical feasibility of digitising large volumes of data from large scale maps.

Section 4

Sample maps by automatic cartography

Section 4
Sample maps by automatic cartography

Notes on the maps

This section of maps is not a "mini-atlas" of a Norwich suburb, but a sample of cartographic methods that might be applied to some of the kinds of data available. There is little significance in the precise selection of area, nor in the actual information which is set out. There is significance, however, in the ability of the system to handle substantial changes of map scale, demonstrated here to a factor of 25 times (25" to 1 mile to 1" to 1 mile). There is also significance in the speeds of output (details of which are given); and in the precision of the output. And finally, there is significance in the technique of 'verbal interrogation' and in the list of questions about this area which the computer was directly able to answer by automatic map drawing.

The subjects which have been handled in these sample maps - Land Use, Rateable Value and Population Characteristics - are rather few, but they make some claim to being typical. In particular, it seemed important to deal in detail with Rateable Value, not simply as a significant subject in its own right, but also to demonstrate the cartographic possibilities that emerge when a subject is handled from geographically detailed raw data (in this case individual hereditaments) which have not been aggregated into districts etc.

It is a matter of regret that these maps have not dealt with any flow problems; the difficulty has been a lack of data, since most of the sampling points for the Norwich Traffic Survey lay outside our chosen 6 kilometre squares. We had also originally hoped to handle some information on landscape or townscape, but here again the field work for this urban area was not available to any useful degree of resolution. Employment details are of obvious planning interest, but these relate to areas topologically similar to, though topographically much smaller in scale than, the Census Enumeration Districts which have been mapped. It will be evident that there are many other parameters which it would have been interesting to map even for this small area, but which had to be excluded on grounds of time and cost.

The maps, in their main groups, are accompanied by notes briefly giving the sources of the data, the editing involved and the lengths of time for the main processes. Some more general notes about the maps section are given below; the techniques are explained more fully in Section 5. Automatic cartography is a developing subject and in the interval between the submission of this report and its passing for press some new techniques have materialised.

Reference has been made above to the need for an automated cartographic system to produce planning information of the same quality of geographical and graphic precision as the base maps on to which it is overlaid. The need for a balanced relationship between the base map (mostly here Ordnance

Survey maps) and the special information is fundamental. The special information must not obscure the base map, nor vice-versa; it is of the essence that the sum of the two should be greater than that of the separate parts. It follows, therefore, that there are legibility problems as between the overlay of new special information and the underlay of topography.

This relationship probably assumes a fairly standard Ordnance Survey base showing a traditional view of topography. It speaks volumes for the good sense of this original definition of topography that it is still so widely accepted as 'basic' today. On the other hand, there may well be elements of the topography on the base map which really are only marginally relevant to planning needs and to the overlaid special information, and which, in this sense, confuse the issue, and contribute to distraction of the eye, if not directly to illegibility. Thus the ability of an automatic cartography system to select what features are to be shown, and in what graphic form, should also be extended to the base map.

Attention must also be directed to the well worn question of the outdatedness of topographic base maps (indeed there is a clearly visible example of a new housing estate within the sample area that is not shown on the Ordnance Survey maps). Maps are always out of date as soon as they are published. However, air photographs do begin to affect this issue especially since they are relatively cheap, and since there are good possibilities of obtaining them for most areas through private organisations or through the Ministry of Housing and Local Government. (Note that a current US Geological Survey estimate suggests that the cost of air photography amounts to only 3% of the cost of topographical map making: special purpose photography, it is concluded, should therefore always be flown when there is a need for information.) There have recently been interesting developments in the technique known as Orthophotography: this technique produces an air photo mosaic corrected to a map-like precision of planimetric detail by the automatic elimination of distortions due to aircraft tilt and to land relief. It can then be overprinted with other information (see Marvin B. Scher: Map use in the District of Columbia – USGS 1969). If such documents can be produced ad hoc, rapidly and cheaply, and to the photographic resolution that the planner requires, they may well form an alternative base to the Ordnance Survey map, and a natural underlay to the special information that can be output by automatic cartography. With the best will in the world, however, orthophotography is unlikely to displace Ordnance Survey mapping for most purposes and it has its own acute problems of legibility; nevertheless, for certain immediate planning uses and in conjunction with automatic cartography, the process could be a very useful one. (Note that the mosaic on page 70 of the map section is not an orthophotograph, and has been made manually from a set of some 12 photographs containing scale errors.)

Automatic cartography offers a repertoire of graphic output, only some of which can be illustrated in a printed report. What are illustrated by the majority of the sample maps are techniques of scribing lines and symbols (to a line gauge of 0.1 mm). This process uses a sapphire point to engrave a white line on an orange-coated scribe sheet — this then requires photographic contacting to positive form in order to be ready for subsequent colour proofing. An alternative method is the use of a Rapidograph pen in place of the scriber. Such pens produce an immediately visible line, though one that is rougher in quality than scribed lines, and they can be filled with inks of various colours. (They need using with some caution as there is a tendency for the ink flow to dry up in the pen and for lines unsuspectedly to disappear.) Apart from these two methods of automatic output, a new light spot projector has recently (July 1970) been fitted to the Unit's drawing machine; this device draws lines with a beam of light on photographic material, and a range of symbols can be produced.

To convert any of the materials produced from the high accuracy plotting table to colour proofs or printed material (as in the map section) some simple cartographic processing is required. This report has not described these well understood techniques of cartographic colour preparation and proofing; but it should be added that they do not necessarily involve complex technical problems or much by way of expensive, special-purpose equipment — a mere drop in the bucket compared with the costs of computer equipment, but an important one nevertheless.

Quite apart from these methods of drawing on the high accuracy plotting table, there are some possibilities in displaying planning information on cathode ray tubes (CRT's). Here again the Unit's storage tube CRT has recently been commissioned (May 1970) and a photograph of its work may be seen in Section 5. However, there are advantages and disadvantages in this type of television-like display; the advantages lie in the speed of display, and in the 'zoom-like' ease with which scales can be changed — both useful qualities in answering questions. The disadvantages lie, first, in the single colour nature of the display and hence the problems of presenting it in relation to base map detail; secondly, in the small size (8" x 5") and in the poor resolution of CRT displays in comparison with printed maps — i.e. scales have generally to be larger to communicate the same amount of information. Finally, of course, a CRT map is inevitably ephemeral in contrast to the permanance of a printed one, and this is not always an advantage.

An alternative method of handling statistical mapping, and one much more in vogue in the USA, is the technique known as SYMAP. This, and a somewhat similar line printer mapping technique called LINMAP, are described in Section 6. Some maps produced by this system for the sample are and from the same data have been included in the map section. It will be appreciated that, while these displays are very quick to produce, their geographical resolution is low, and since there is little incentive to overlay them on Ordnance Survey base maps, they tend to stand alone. A graphically more versatile system has been developed within the Unit by employing the Photon 540 computer typesetter at the London College of Printing. An example of this technique used on Population Census data has been included in Section 5.

Mention must also be made of Overhead Projectors as a means of presenting planning maps. Not only does the overhead projector enable a speaker to communicate cartographic problems to a large audience (as in a teacher-class relationship) but, more significantly, it enables him to interchange and select from a whole range of overlays so as to show the effects of superimposing one parameter on another — and this is something that is difficult to do with normal printed maps. Overhead projectors (which cost between £100 and £1,000) do seem to have possibilities as cartographic research tools, and are in use as such by at least one progressive planning authority.

In addition to producing maps it is obviously desirable for any automatic system to be able to supply information from the databank in the form of alphabetical listings, directories and the like. The relationship between lists of Rateable Values and the mapping of these has been referred to above. This study did not include work on gazetteer or street directory lists produced from the computer on a line printer or automatic type-set basis.

Too often the normal map, if asked about a specific question of detail, can give only a general indication of the answer. The particular enquiry can only be met by hunting through the raw data, then plotting the information ad hoc on the general map. This situation is capable of very great improvement by a computerised cartographic system where the raw data are held on magnetic tape. In this particular study the following detailed questions — about rateable values — were asked of the computer (through the teletype) and the answers were then plotted by symbols direct on to a tracing paper overlay at 6 ins to the mile scale, and within a time of 30 seconds to 5 minutes, depending on the amount of detail to be drawn.

"Which houses have a rateable value between £40 and £60; which have been built since 1963; which have been taken out of the rating records since 1963?
"Which hereditaments have gone down in value by £10 or less between 1963 and 1969; which have gone up by £50 or more?
"What is the total rateable value of all hereditaments inside this area (which may be a polygon of any shape provided that any vertical line cuts the outline twice only)?
"What is the total rateable value of all the hereditaments in each postal code; which postal codes fall between £1,000 - £2,000 in aggregated rateable value?
"Which grid squares have a rateable value over £2,000; which have a mean value under £50?
"What are the numbers of hereditaments falling in each £10 interval between £0 - £200?
"What are the number of grid squares falling in each £50 interval between £0 - £4,000?"

Sample maps

Subject	**Land Use**
Scale	6″ = 1 mile (1:10,560)
Data	The land use categories and boundaries were derived from the maps produced by the planning departments of Norwich City and Norfolk County in connection with the 1967 Norwich Area Land Use/Transportation Survey.
Pre-input editing	The 1967 Survey involved 19 use categories which, for this project, were amalgamated to form the basic 9 groups. The categories were transferred onto tracing paper overlays at 1:2,500 scale (6 sheets) and this process took 13 man hours.
Digitising	Digitising for all Land Use Maps was at 25″ to the mile scale (6 sheets); time taken was 26½ man hours approximately.
Correction	Each segment of boundary was checked on-line against its feature code and any necessary corrections were made. For the 6 sheets, the time taken was approximately 6 hours.
Output	The boundaries and symbols were scribed by the geagraph on separate sheets from which colour masks were produced direct. Drawing time: 146 minutes (of which the symbols took 27 mins., the total pattern line 42 mins., and the "residential + hotels" category 23 mins.).

Legend

+	Shops, offices and petrol filling stations
Δ	Commerce, public transport (e.g. railways), car-parking
×	Residential and hotels
□	Open space (public and private), woodland
★	Public buildings (education, health, library, etc.)
▽	Industry, minerals, building sites and public utilities
◇	Rural (but excluding woodland)

Note: Water and roads have been left blank.

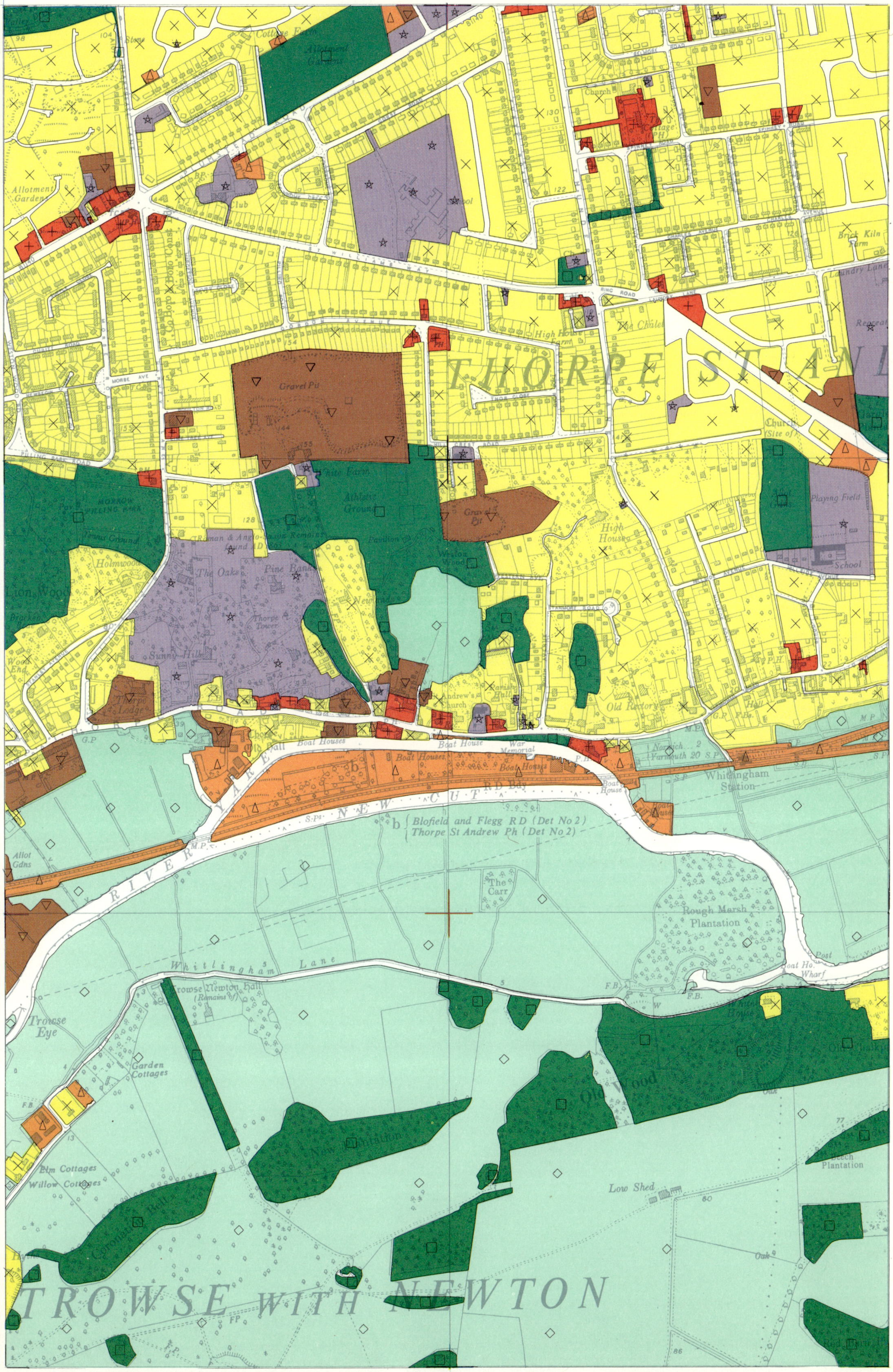
THORPE ST ANDREW
TROWSE WITH NEWTON
Allotment Gardens
Gravel Pit
Athletic Ground
Gravel Pit
High House
Playing Field
School
The Oaks
Pine Banks
Thorpe Tower
Sunny Hill
Thorpe Lodge
Lion Wood
Old Rectory
Boat Houses
Boat House
War Memorial
Whitlingham Station
RIVER YARE
NEW CUT
Blofield and Flegg R D (Det No 2)
Thorpe St Andrew Ph (Det No 2)
The Carr
Rough Marsh Plantation
Whitlingham Lane
Trowse Newton Hall (Remains of)
Trowse Eye
Garden Cottages
Elm Cottages
Willow Cottages
Coronation Belt
New Plantation
Old Wood
Low Shed
Beech Plantation
Brick Kiln Farm
Laundry Lane
The Chalet
Church (Site of)
Allotment Gardens
Allot Gdns
Norwich 2
Yarmouth 20

Subject	**Land Use**
Scale	25″ = 1 mile (1 : 2,500)
Data Pre-input editing Digitising Correction	See notes on page 40
Output	The boundaries and symbols were drawn using a rapidograph pen with the geagraph. Drawing time: 10 minutes (of which the pattern line took 6 mins.).

Legend

+ Shops, offices and petrol filling stations

∆ Commerce, public transport (e.g. railways), car parking

✕ Residential and hotels

□ Open space (public and private), woodland

★ Public buildings (education, health, library, etc.)

∇ Industry, minerals, building sites and public utilities

◇ Rural (but excluding woodland)

Note: Water and roads have been left blank

WESTON WOOD ROAD
WESTON WOOD CLOSE
Red Roofs
Gravel Pit
High House
THUNDER LANE
BM 83·09
Weston Wood
Ruin
WESTERN AVENUE
Tennis Court
SOUTH AVENUE
STANMORE ROAD
Pavilion
Tennis Courts
St Andrew's Church (C of E)
Rectory
St Andrew's Church
Church Cottages
Hill Cottage
Garage
Ramp
1 to 16
Richmond Court
YARMOUTH ROAD
BM 9·84
BM 10·91
War Memorial
RIVER GREEN
2238 ·02
RIVER YARE
2434 ·01
2634 ·01
1934 ·06
1532 3·20
1728 1·90
0032 ·18
Co Const & R D Bdy
C B & Boro Const Bdy
Thorpe Gardens (P H)
Depot
Works
Boat House
High Water Mark of Medium Tides
Level Crossing
Drain
3629 ·08
1827 1·47
R D Bdy
NEW CUT
CCLW

Subject	**Land Use**
Scale	2½″ = 1 mile (1 : 25,000)
Data Pre-input editing Digitising Correction	See notes on page 40
Output	The boundaries and symbols were scribed by the geagraph. Drawing time: Pattern line 30 minutes, Symbols 19 minutes, total 49 minutes.

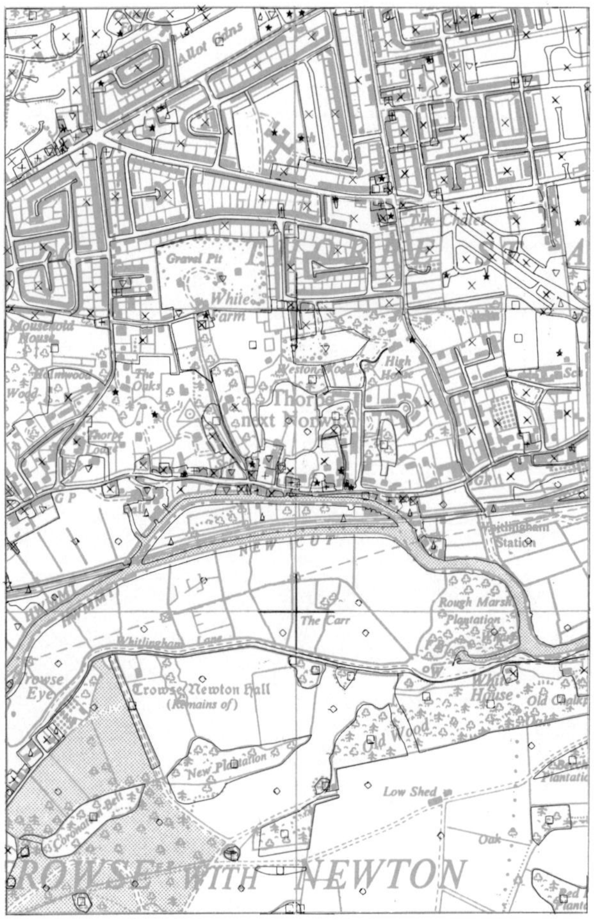

Legend

- \+ Shops, offices and petrol filling stations
- △ Commerce, public transport (e.g. railways) car parking
- ╳ Residential and hotels
- □ Open space (public and private), woodland
- ✩ Public buildings (education, health, library, etc.)
- ▽ Industry, minerals, building sites and public utilities
- ◇ Rural (but excluding woodland)

Note: Water and roads have been left blank

Subject	**Land Use**
Scale	1″ = 1 mile (1 : 63,360)
Data Pre-input editing Digitising Correction	See notes on page 40
Output	The basic data are the same as have been used in the last 3 maps, but in this case the seven categories have been grouped to form only four. The computer selected the relevant boundaries and the outlines have been slightly generalised at this smaller scale (i.e. drawn with larger plotting tolerance). The boundaries were scribed by the geagraph separately for each map. Drawing time: Map (a) 7 minutes, (b) 7 minutes, (c) 11 minutes, (d) 7 minutes, total pattern line, 20 minutes.

a

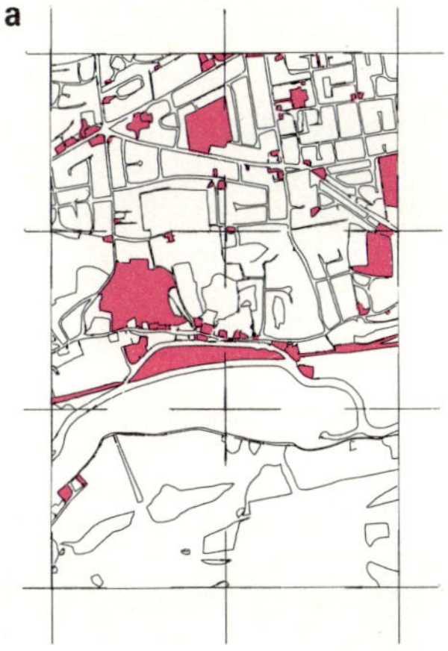

a Shops, offices, petrol filling stations, commerce, public transport, car parks public buildings.

b Open space and rural

b

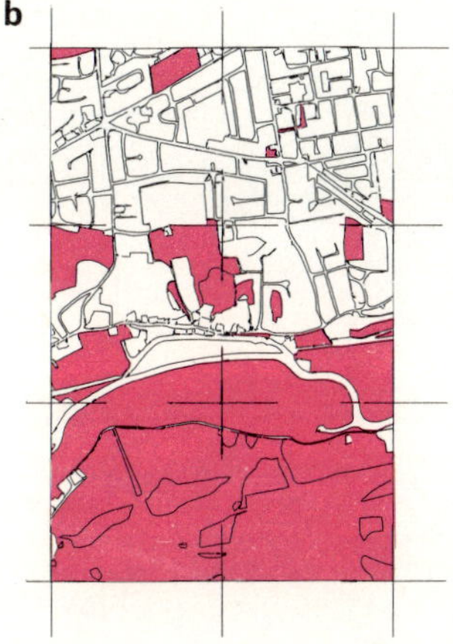

c

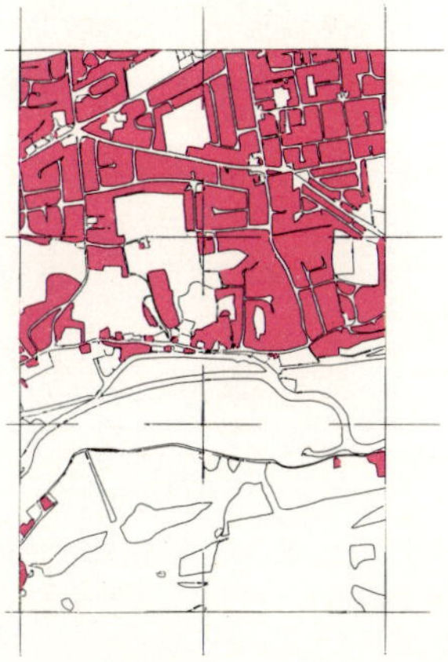

c Residential and hotels

d Industry, minerals, building sites and public utilities

d

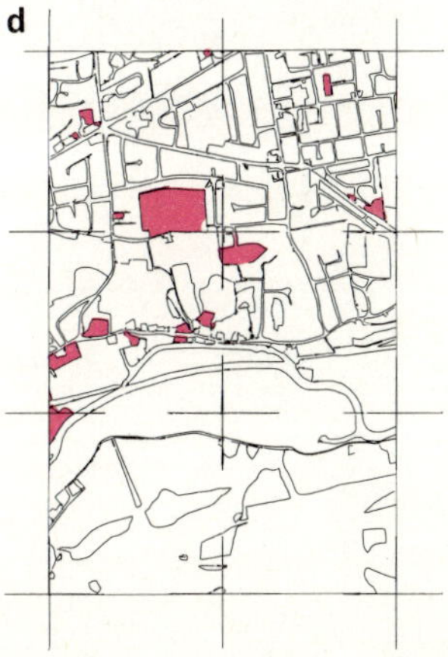

Subject	**Rateable Value 1969; by individual hereditament**
Scale	25" = 1 mile (1 : 2,500)
Data	The rateable value data for individual hereditaments are held by the Local Authority for rate collection purposes, and also by the Inland Revenue whose job it is to assess the values and apportion the rates between different categories of hereditament.
Pre-input editing	The figures for the net rateable value assessment per hereditament in 1963 (the last major revaluation) were written on OS base maps at 50" (16 sheets) and 25" (2 sheets) to 1 mile scale. The 1963 figure applied also in 1969 if no change had occurred; in cases where the value had changed, the 1969 figure was also written onto the base maps.
Digitising	Digitising was in street sequence within sheets at 50" to 1 mile scale (16 sheets) and 25" (2 sheets); time taken was 35 man-hours. This digitised data then applied for all scales of map.
Correction	A listing of the values was produced on a teletype and checked (off-line) with a plot of the positions of hereditaments; corrections were made. This operation took approximately 15 man-hours.
Output	The boundaries and symbols were scribed by the geagraph in approximately 8 minutes. The wedges of half the map have been peeled to appear solid.

Legend

- £ 1 - 39
- £ 40 - 59
- £ 60 - 99
- £ 100 - 159
- £ 160 +

Note: The symbols in the SE half of the map remain open - an alternative method of presenting the same information. In cases of overlap (e.g. Thunder Lane in the east) the open symbols are less ambiguous than solid ones would have been, and they do not mask base map detail; in general they are less easily seen than the solid black wedges.

Weston Wood Close
Red Roofs
BM 138·33
Disused Gravel Pit
High House
Thunder Lane
Weston Wood
Ruin
BM 83·09
Western Avenue
Tennis Court
Stanmore Road
South Avenue
Stanmore Close
Pavilion
Tennis Courts
Chapel Lane
Hall
Hill Cottage
St Andrew's Church (C of E)
Church Cottages
Rectory
St Andrew's Church
Buck Yard
Garage
PO
Ramp
Richmond Court
Yarmouth Road
BM 9·84
War Memorial
River Green
TCB
2238
·02
BM 10·91
River Yare
2434
·01
2634
·01
Co Const & R D Bdy
CB & Boro Const Bdy
Thorpe Gardens (PH)
Depot
1934
·06
1532
3·20
1728
1·90
0032
·18
Works
Boat House
High Water Mark of Medium Tides
Drain
Level Crossing

Subject	**Rateable Value 1969; 5 classes for individual hereditaments**
Scale	6″ = 1 mile (1 : 10,560)
Data	As for map on page 47 (see notes on page 46) with the addition of the outline of an enumeration district, obtained from the General Register Office.
Pre-input editing Correction	See notes on page 46
Digitising	The outline of the enumeration district was digitised at the scale of 6″ to 1 mile. See notes on page 46 for detail of rateable value digitising.
Output	Five classes of rateable value were established (see Legend) and a separate colour mask was cut for each category. The symbols are small squares scribed by the geagraph. Drawing time: 118 minutes (of which the £60 - 99 category took 59 minutes).

The outline of Enumeration District number 140703 has been superimposed on the map. The total rateable value within this area was calculated by the computer from the original records (digitised in street order) and is £76,456 for 1969. The time to calculate this value (including digitising) was 7 minutes.

Note: The congested kilometre square from the NW corner is reproduced here with a smaller symbol demonstrating the possibilities of dealing with dense urban situations whilst still retaining detail for individual houses.

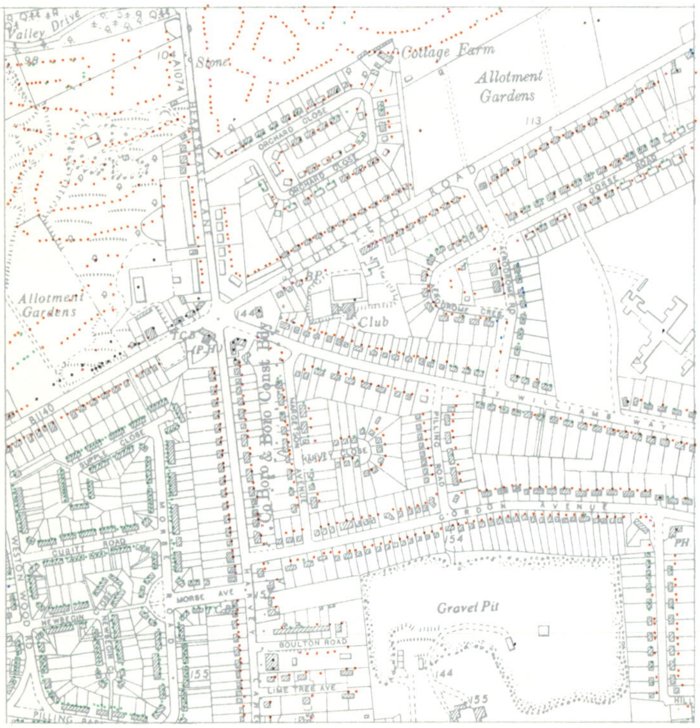

Legend

Blue	£ 1 - 39
Green	£40 - 59
Orange	£60 - 99
Red	£100 - 159
Black	£160 +

THORPE ST AND
TROWSE WITH NEWTON
Valley Drive
Cottage Farm
Allotment Gardens
Church
The Cottage (P.H.)
School
Club
Brick Kiln Farm
Laundry Lane
Ring Road
The Chalet
Recreat
Gravel Pit
Boulton Road
Lime Tree Ave
Thor Close
Hill Crest Road
Church (Site of)
Gargle Plant
White Farm
Athletic Ground
Morrow Pilling Park
Pav.
Tennis Ground
Roman & Anglo-Saxon Remains found AD 1863
Pavilion
Gravel Pit
Weston Wood
Tower
High House
Allot. Gdns.
Playing Field
School
Holmwood
Lion Wood
The Oaks
Pine Banks
Newlands
Stanmore Road
Bracken Brae
Thorpe Tower
Wood End
Sunny Hill
School
St Andrew's Church
Parish Hall
Old Rectory
Thorpe Lodge
Thorpe Old Hall
Boat Houses
Boat House
War Memorial
Boat Houses
Norwich 2
Yarmouth 20
Whitlingham Station
Boat House
River Yare
New Cut
RD Bdy
b
Blofield and Flegg R D (Det No 2)
Thorpe St Andrew Ph (Det No 2)
Allot Gdns
The Carr
Rough Marsh Plantation
Boat Ho. Wharf
Post
Whitlingham Lane
Trowse Newton Hall (Remains of)
Trowse Eye
White House
Old Chalkp
Garden Cottages
Old Wood
New Plantation
Beech Plantation
Elm Cottages
Willow Cottages
Low Shed
Coronation Belt
Hythe
Oak
Red Barn Pl

Subject	**Rateable Value 1969; totals for 100 metre squares** (shown opposite)
Scale	6″ = 1 mile (1 : 10,560)
Data Pre-input editing Correction	See notes on page 46
Digitising	Digitising for each hereditament was at the assumed centre of each building; there may therefore be some misrepresentation in cases where buildings overlap 100 metre grid lines.
Output	The aggregation is to 100-metre squares and the geagraph scribed the symbols in 23 minutes. The squares were then "peeled" to produce negatives. Mapping total values, it is not possible to work out the significance of the amount of rateable value in isolation from the base map (i.e. the number of hereditaments is crucial).

Legend

£5 - 404 □   £805 - 1204 □   £1605 + □

£405 - 804 □   £1205 - 1604 □

Subject	**Rateable Value 1969; total values by 500 metre squares** (shown below)
Scale	1″ = 1 mile (1 : 63,360)
Data Pre-input editing Digitising Correction	See above
Output	Rateable values aggregated to 500 metre squares with 3 class intervals. The symbols were scribed by the geagraph for the 2 maps in 3 minutes.

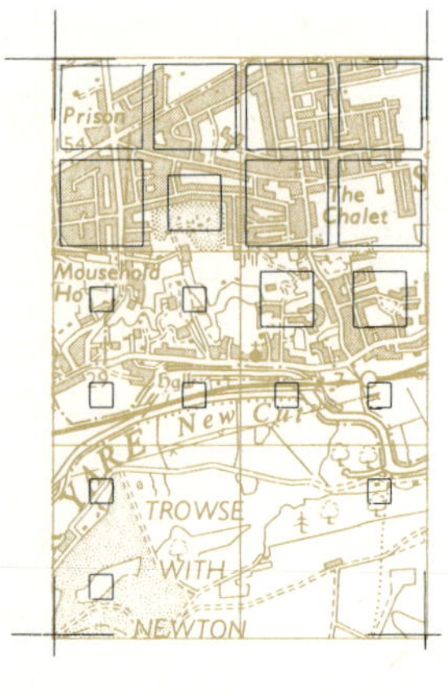

Legend

£ 1 - 10,000 □

£10,001 - 20,000 □

£20,001 + □

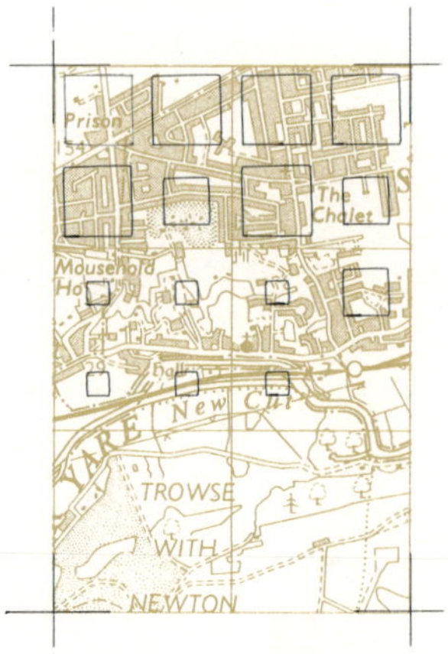

£4001 - 14,000 □

£14,001 - 24,000 □

£24,001 + □

Valley Drive
Stone
Cottage Farm
Allotment Gardens
Church
The Cottage (PH)
Allotment Gardens
Club
School
Brick Kiln Farm
Laundry Lane
Charles Avenue
Gordon Avenue
High House Farm
The Chalet
Recreat
THORPE ST ANI
Gravel Pit
Church (Site of)
Gargle Planta
Thor Close
Lime Tree Ave
Hill Crest Road
Pilling Park Road
White Farm
Athletic Ground
Ch
Allot. Gdns.
Playing Field
Pav.
MORROW PILLING PARK
Tennis Ground
Roman & Anglo-Saxon Remains found AD 1863
Pavilion
Gravel Pit
High House
Weston Wood
Holmwood
The Oaks
Pine Banks
School
Lion Wood
Bracken Brae
Newlands
Western Ave
Stanmore Road
Thorpe Tower
Wood End
Sunny Hill
Thorpe Lodge
School
St Andrew's Church
Parish Hall
Old Rectory
Hall
Thorpe Road
Thorpe Old Hall
Boat Houses
Boat House
War Memorial
Boat Houses
Norwich 2
Yarmouth 20
Whittingham Station
R D Bdy
Boat House
New Cut
River Yare
b
b Blofield and Flegg R D (Det No 2)
Thorpe St Andrew Ph (Det No 2)
Allot Gdns
The Carr
Rough Marsh Plantation
Post
Boat Ho
Wharf
Whitlingham Lane
Trowse Newton Hall (Remains of)
Trowse Eye
White House
Old Chalkp
Garden Cottages
Old Wood
New Plantation
Beech Plantation
Elm Cottages
Willow Cottages
Low Shed
Coronation Belt
Hythe
Oak
TROWSE WITH NEWTON
Red Barn Pl

Subject	**Rateable Value; change 1963/69 by individual hereditament**
Scale	6" = 1 mile (1 : 10,560)
Data	As for map on page 47 (see notes on page 46) with the addition of the outline of a quite arbitrary shape.
Pre-input editing Correction	See notes on page 46
Digitising	The outline of the previously undefined polygon was digitised at the scale of 6" to 1 mile. See notes on page 46 for details of rateable value digitising.
Output	The symbols identify hereditaments which have increased in value (red), decreased in value (blue) and also indicate those which were not in existence in 1963 (black). Drawing time: 52 minutes (of which the symbols showing an increase took 21 minutes).

The irregular polygon (large chevron on facing page) was treated in the same way as the enumeration district (see pages 48/9). The outline was digitised and the necessary calculations for figures within it performed in 7 minutes. The total rateable value for 1969 in this area is £10,422.

Legend

Red	Hereditaments with an RV increase 1963/69
Blue	Hereditaments with an RV decrease 1963/69
Black	Hereditaments not in existence in 1963, but present in 1969 records.

The histograms below show the number of occurrences of hereditaments within £10 groupings of rateable value. The large number in the first column for 1963 represents houses which were new in 1969 and were regarded as having a rateable value of £0 in 1963. Drawing time for the two histograms: 20 minutes.

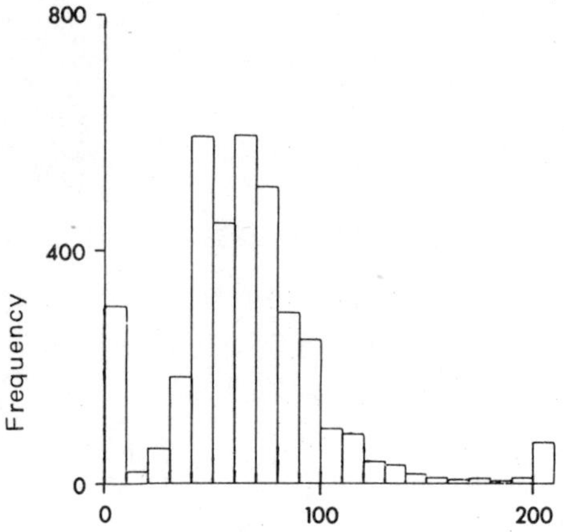

1963 Rateable Values (£)

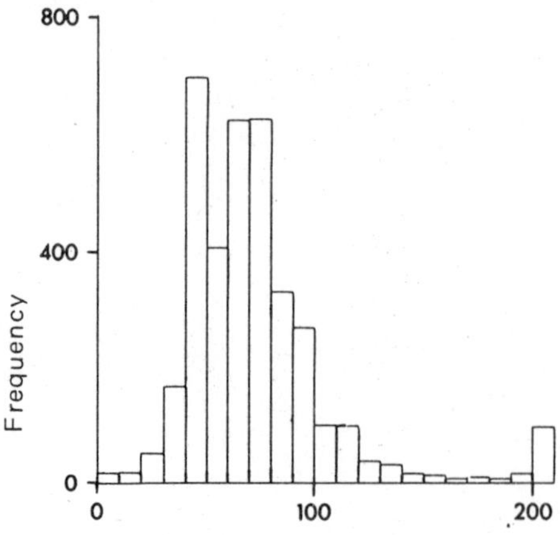

1969 Rateable Values (£)

Note: The final column in each histogram includes all values between £200 and the maximum for this area (£4,500)

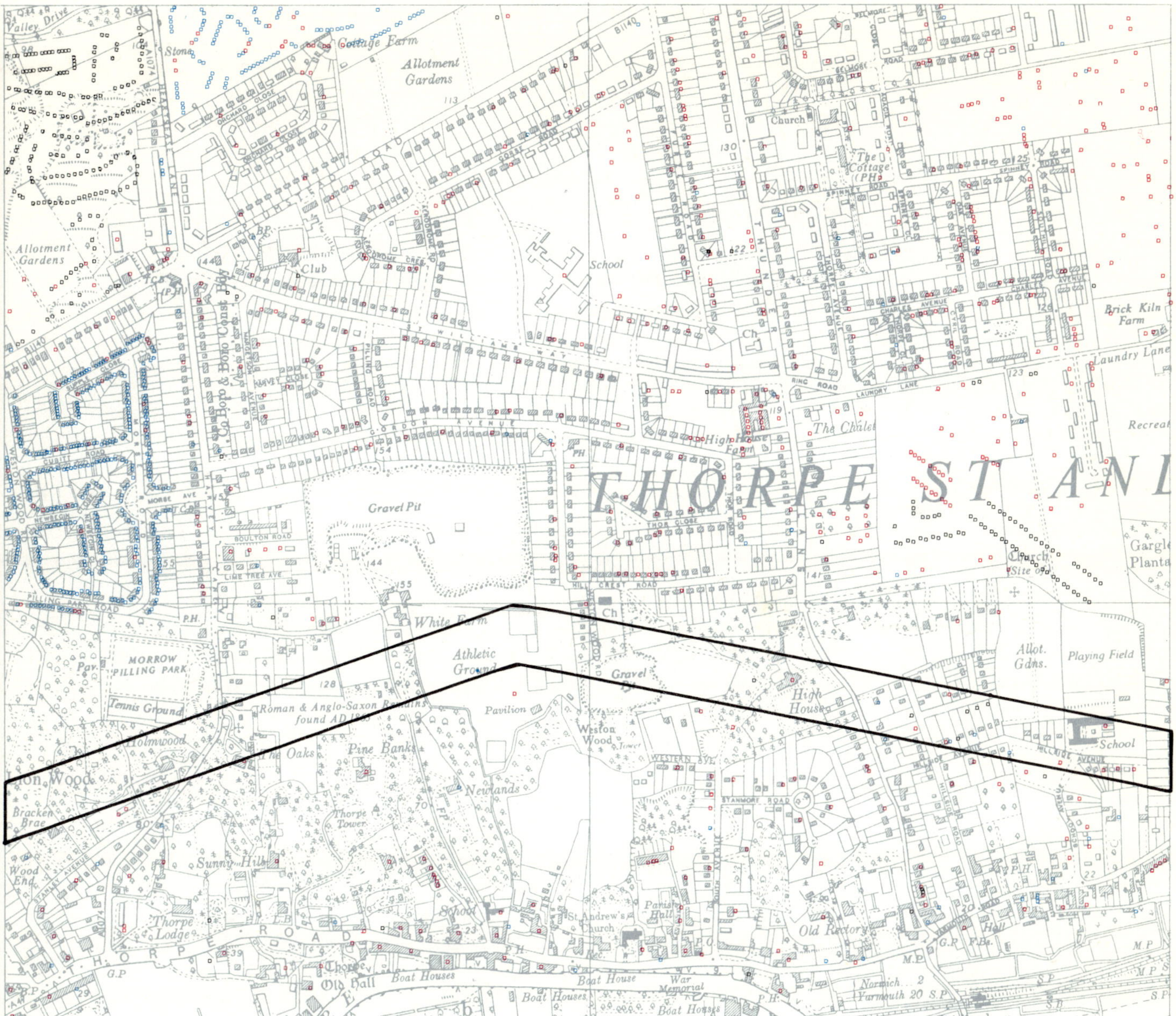

THORPE ST AND
TROWSE WITH NEWTON
Valley Drive
Cottage Farm
Allotment Gardens
Allotment Gardens
School
Church
The Cottage (PH)
Brick Kiln Farm
Laundry Lane
Recreat
Club
Gravel Pit
White Farm
Athletic Ground
Pavilion
Weston Wood
High House Farm
The Chalet
Church Site
Gargle Planta
Allot. Gdns.
Playing Field
School
High House
Morrow Pilling Park
Pav.
Tennis Ground
Roman & Anglo-Saxon Remains found AD 1898
Holmwood
The Oaks
Pine Banks
Newlands
Bracken Brae
Thorpe Tower
Sunny Hill
Wood End
Thorpe Lodge
School
St Andrew's Church
Parish Hall
Old Rectory
Thorpe Road
Thorpe Old Hall
Boat Houses
Boat House
War Memorial
Boat Houses
Norwich 2
Yarmouth 20
Whitlingham Station
River Yare
New Cut
RD Bdy
b
b { Blofield and Flegg RD (Det No 2)
Thorpe St Andrew Ph (Det No 2)
Allot Gdns
The Carr
Rough Marsh Plantation
Whitlingham Lane
Trowse Newton hall (Remains of)
Trowse Eye
Boat Ho. Wharf
White House
Old Chalkp
Garden Cottages
Old Wood
New Plantation
Beech Plantation
Elm Cottages
Willow Cottages
Low Shed
Coronation Belt
Hythe
Oak
Red Barn Pl

Subject	**Rateable Value 1969; mean values for 100 metre grid squares using SYMAP technique - choropleth map**
Scale	2½" = 1 mile (1 : 25,000, photo-reduced from 1:10,560)
Data Pre-input editing Digitising Correction	See notes on page 58 and page 46
Output	Map based on data from all 600 100-metre squares; the lowest level has been chosen to indicate areas where data do not exist (i.e. no buildings to be rated). For gridded data, the proximal map (print position values are those of the nearest data point) output should be identical to that produced by choropleth techniques. This obviates the need to produce 5 cards containing the co-ordinates of the four corners and a terminator for each data area, a necessity in the standard SYMAP choropleth technique. Six levels on an arithmetic scale have been employed and are represented by the standard lineprinter SYMAP symbolism. Production time: 1 . 90 minutes on an IBM 360/50 at the Edinburgh Regional Computing Centre.

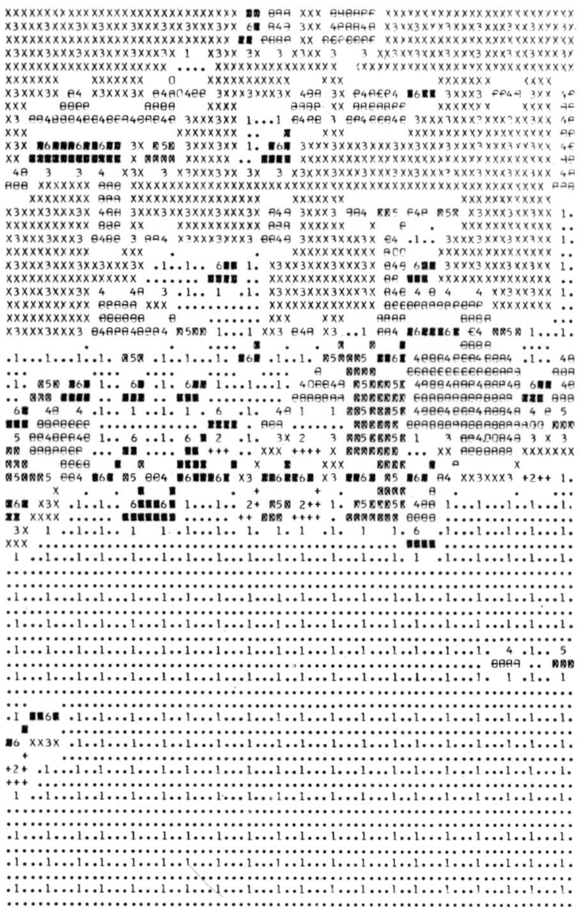

Legend

Level	Value	Representation
1	£0 - 1	...1...
2	£2 - 40	+++2+++
3	£41 - 80	XX3XX
4	£81 - 120	θθθ4θθθ
5	£121 - 160	▩▩5▩▩
6	£161 - 5,000	■■6■■

Subject	**Rateable Value 1969; mean values for 100 metre grid squares using SYMAP technique - contour map**
Scale	6″ = 1 mile (1 : 10,560)

Data, Pre-input editing, Digitising, Correction: See notes on page 58 and page 46

Output

The contour map is based on the values in grid squares where property exists to be rated: thus 290 data points were employed, the 310 squares containing no rateable property being filtered out in the Flexin sub-routine.

Six levels on a geometric scale have been specified and are reproduced by non-standard line printer symbolism (see Legend). Numerical values indicate the position of a data point whose value lies within the level given by the digit. Production time: 1.05 minutes on an IBM 360/50 at the Edinburgh Regional Computing Centre.

Legend

Level	Value	Representation
1	£0 - 15	---1---
2	£16 - 35	'''2''''
3	£36 - 75	III3III
4	£76 - 155	0004000
5	£156 - 315	XXX5XX
6	£316 - 5,000	6

Subject	**Rateable Value 1969; aggregated by postal codes**
Scale	Page 56: 2½" = 1 mile (1 : 25,000) Page 57: 6" = 1 mile (1 : 10,560)
Data Digitising Correction	See notes on page 46
Pre-input editing	Hereditaments were coded in groups by street and by postal code. This made aggregation of rateable value to street or postal code possible.
Output	Rateable values were aggregated within postal codes and these totals were grouped into 5 classes (see Legend). In the map on this page (page 56), lines have been drawn by the geagraph with a rapidograph pen joining hereditaments within postal codes. Drawing time: c. 30 minutes.

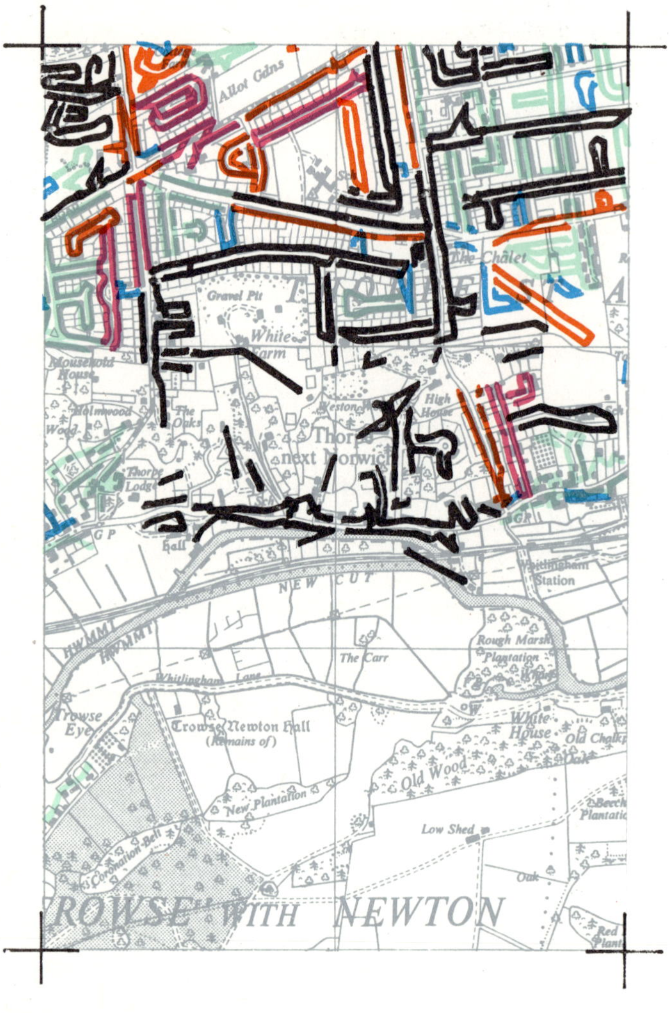

Legend

Blue	£ 1 - 1,500
Green	£1,501 - 3,000
Orange	£3,001 - 4,500
Red	£4,501 - 6,000
Black	£6,001 +

On this map, the hereditaments are identified separately by symbols scribed by the geagraph in 152 minutes.

Legend

Black	£ 1 - 1,500
Red	£1,501 - 3,000
Orange	£3,001 - 4,500
Green	£4,501 - 6,000
Blue	£6,001 +

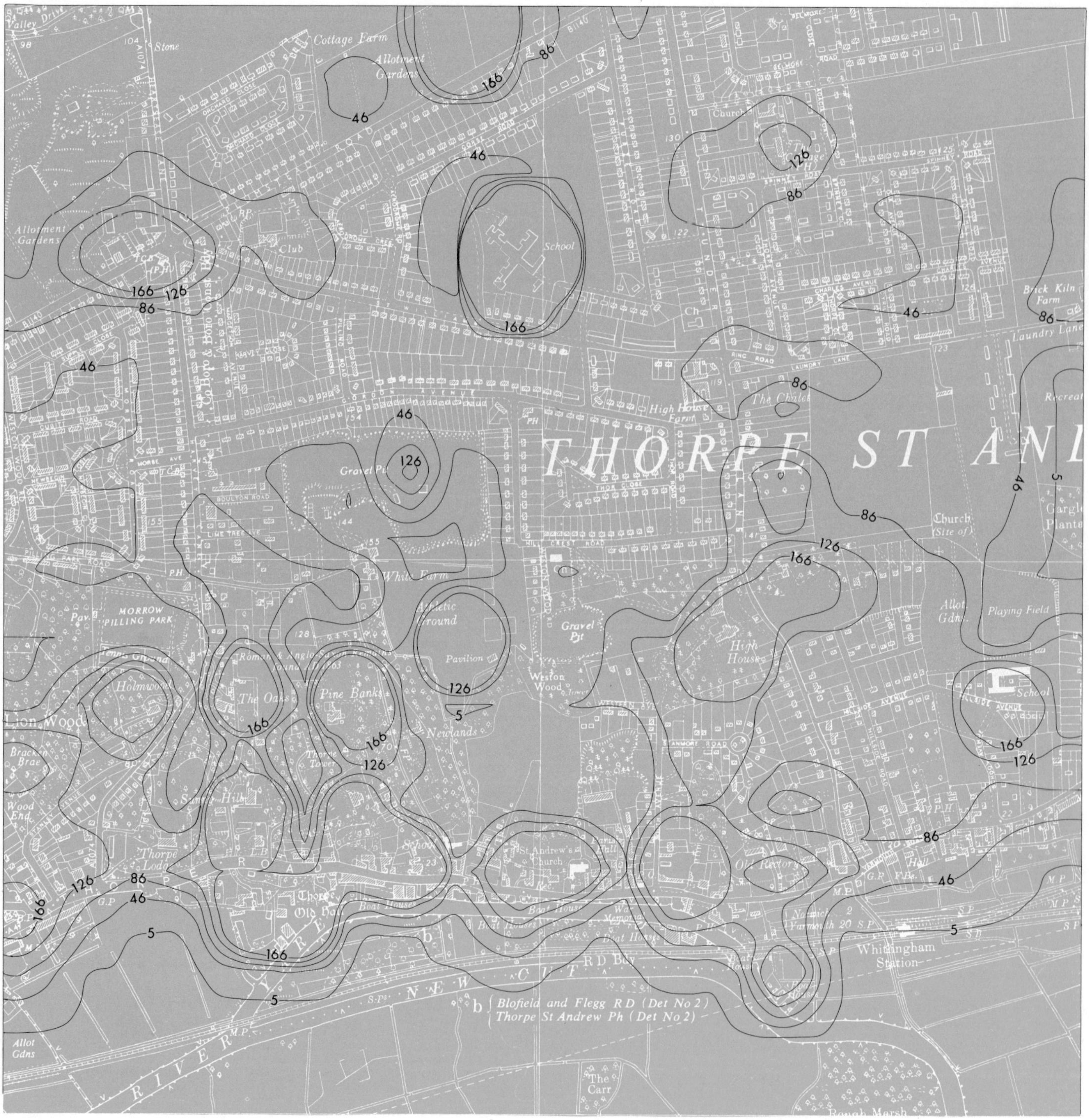

Subject	**Rateable Value 1969; contoured from means for 100 metre squares**
Scale	6″ = 1 mile (1 : 10,560)
Data	Total rateable value for each square was divided by the number of contributory hereditaments, giving a mean value per hereditament.
Pre-input editing Digitising Correction	See notes on page 46
Output	The technique is exactly the same as for map on page 59 of contours for grid totals. Contour levels: £ 5, 46, 86, 126, 166. Drawing time: 35 minutes. (Numbers inserted by hand.)

Subject	**Rateable Value 1969; contoured from totals for 100 metre grid squares**
Scale	6″ = 1 mile (1 : 10,560)
Data	The 100-metre grid square totals (see pages 50/51) were used as the basis for the contouring.
Pre-input editing Digitising Correction	See notes on page 46
Output	Using the rateable value grid totals as its figure field, a program interpolated the contours at specified levels (£ 5, 405, 805, 1205, 1605, 2005). The contour lines were scribed by the geagraph in 37 minutes. Note: This technique might be applied to confidential data since it presents the patterns and distributions without revealing the detailed data.

Subject	**Rateable Value 1969; anaglyph version of contours from totals for 100 metre squares (page 59)**
Scale	6″ = 1 mile (1 : 10,560)
Data Pre-input editing Digitising Correction	Exactly as map on page 59
Output	The map on page 59 is produced here in anaglyph form. When viewed through red and green spectacles, the impression gained is three-dimensional. The computer can produce this type of presentation which involves drawing each contour twice, once in red and once in green. Each red contour is displaced to the left and each green a corresponding distance to the right, the amount of displacement being proportional to the actual (or, in this case, conceptual) height of each contour.

Subject	**Rateable Value 1969; surfaces presented as isometric block diagrams**
Scale	The vertical factor (see maps) is the relationship between £'s and millimetres, e.g. if at a particular point the value is, say, £1,000, with a vertical factor of 0.004, this would be represented by 4 millimetres. Scale along grid lines: 2½" = 1 mile (1 : 25,000)
Data Pre-input editing Digitising Correction	The diagrams are based on the same information as used in the maps on pages 50 and 51 but the program manipulates the 20 x 30 matrix into one with 21 x 31 data points (i.e. positioned at the intersections of grid lines rather than the centres of squares).
Output	Each of the diagrams was drawn by the geagraph in roughly 20 minutes. Two were scribed and the third drawn by rapidograph pen. The lines in the diagrams represent the 100-metre grid lines. Note that the "hidden" lines have been removed by hand.

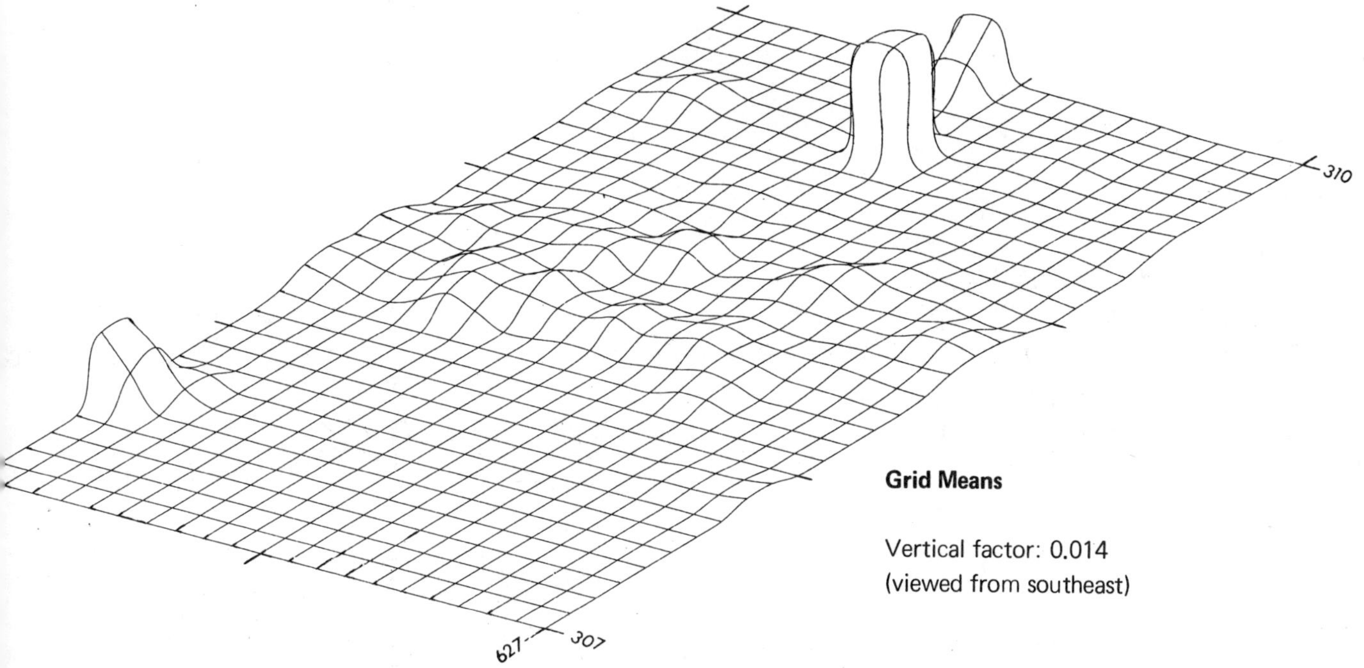

Grid Means

Vertical factor: 0.014
(viewed from southeast)

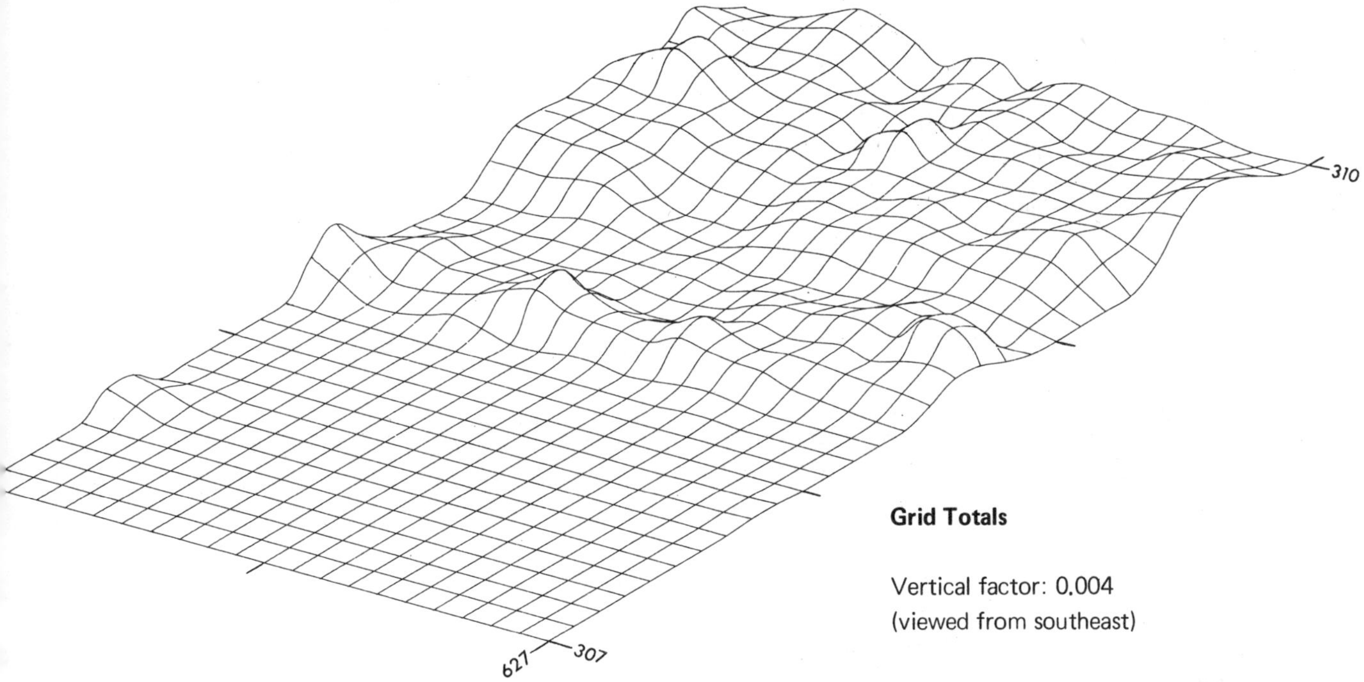

Grid Totals

Vertical factor: 0.004
(viewed from southeast)

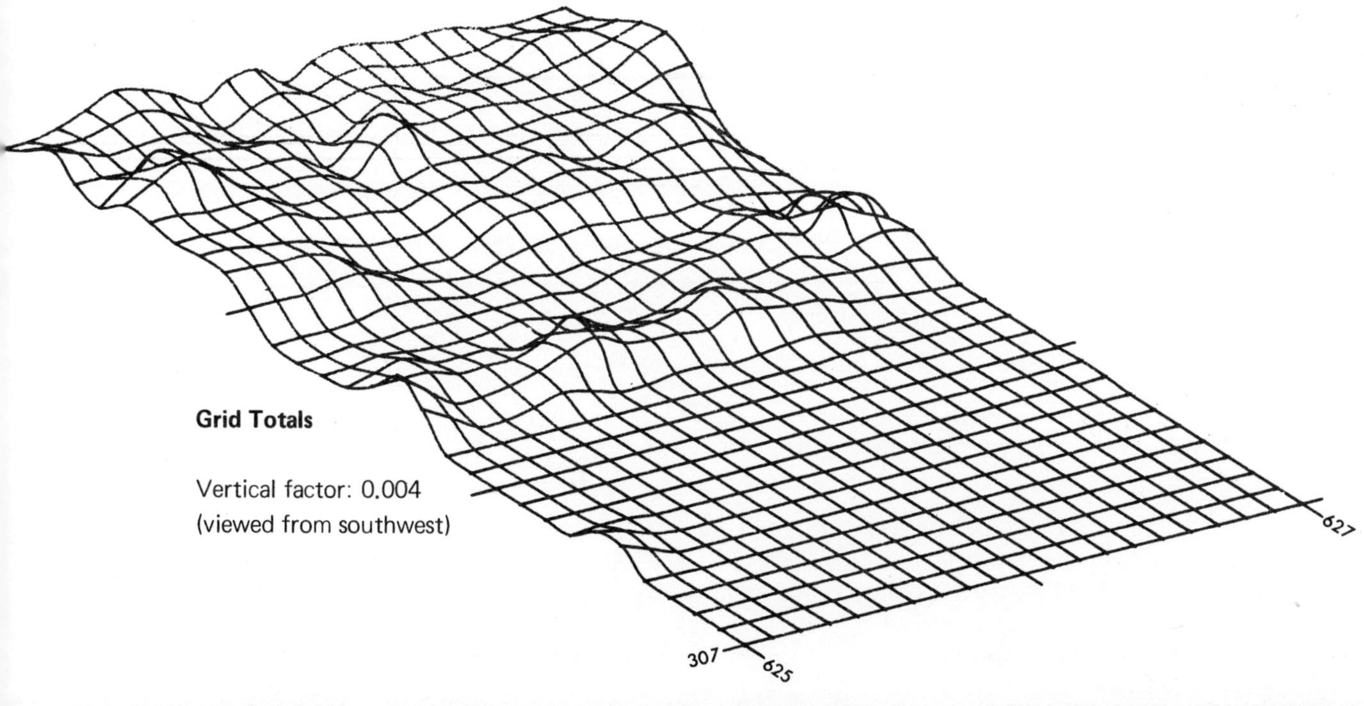

Grid Totals

Vertical factor: 0.004
(viewed from southwest)

Subject	**Rateable Value; change between 1963 and 1969 by 100 metre square**
Scale	6″ = 1 mile (1 : 10,560)
Data Pre-input editing Digitising Correction	As for pages 50/1, with the addition of data for 1963.
Output	The squares represent class intervals of data (see Legend). 1969 values (in red) were drawn by pen on one sheet and 1963 values on another (in blue). Where there has been no change from 1963, the coloured squares are superimposed. (NB. There may have been change within a class interval which would not show with this particular group of classes). Drawing time: 42 minutes.

Legend

£ 1 - 400

£401 - 800

£801 - 1,200

£1,201 - 1,600

£1,601 +

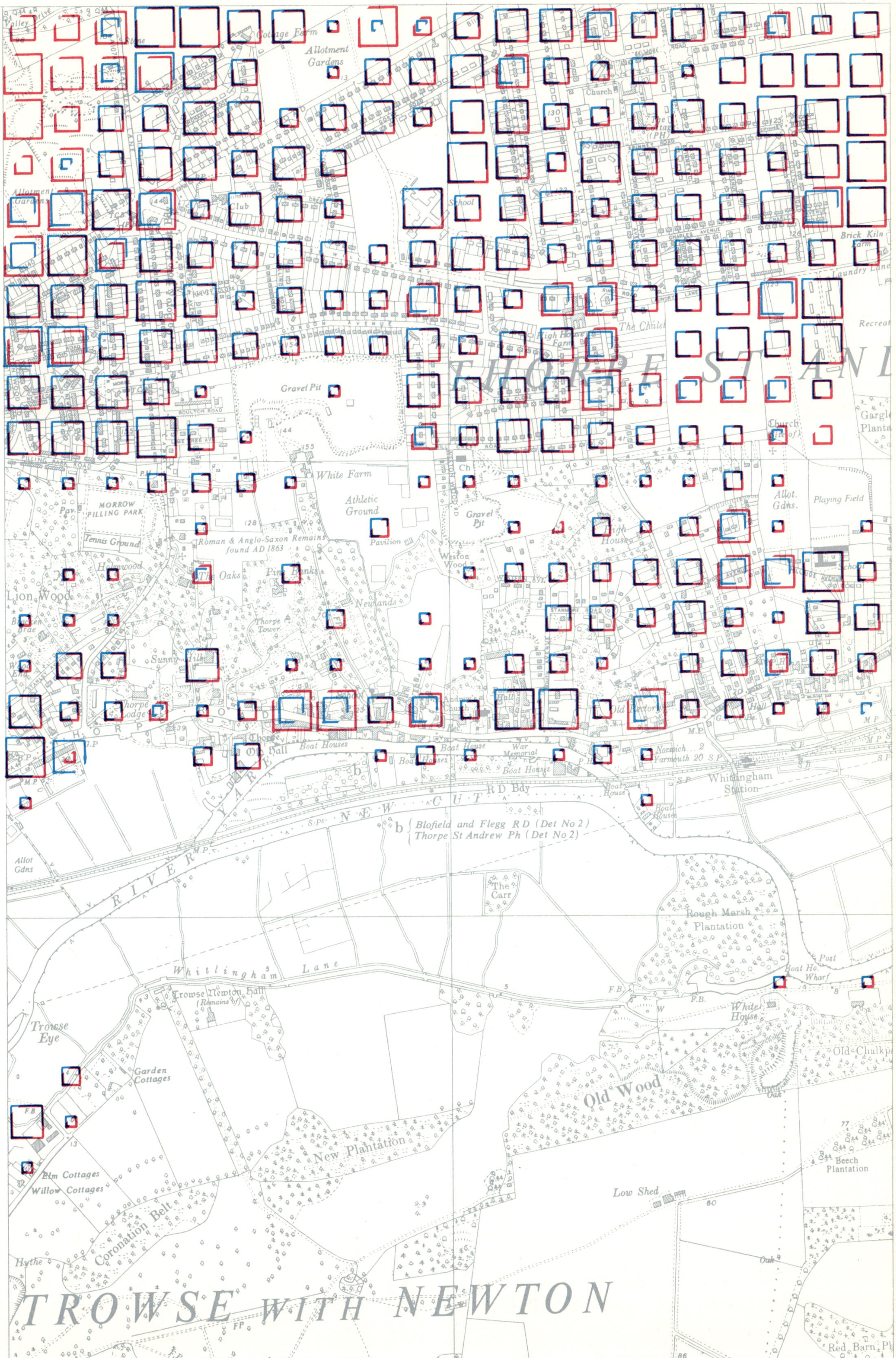
Cottage Farm
Allotment Gardens
Church
School
Club
Brick Kiln Farm
Laundry Lane
The Chalet
High House Farm
THORPE ST AND
Gravel Pit
Gargle Plantation
White Farm
Athletic Ground
Pavilion
Weston Wood
Allot. Gdns.
Playing Field
MORROW PILLING PARK
Tennis Ground
PILLING PARK ROAD
BOULTON ROAD
GORDON AVENUE
CHARLES AVENUE
Roman & Anglo-Saxon Remains found AD 1863
The Oaks
Pine Banks
Newlands
Lion Wood
Thorpe Tower
Sunny Hill
Thorpe Lodge
Old Rectory
Hall
Boat Houses
Boat House
War Memorial
Whitlingham Station
Norwich 2
Yarmouth 20
RD Bdy
NEW CUT
RIVER YARE
Blofield and Flegg RD (Det No 2)
Thorpe St Andrew Ph (Det No 2)
Allot Gdns
The Carr
Rough Marsh Plantation
Whitlingham Lane
Trowse Newton Hall (Remains of)
Trowse Eye
Garden Cottages
Elm Cottages
Willow Cottages
New Plantation
Old Wood
Old Chalkpit
White House
Boat Ho.
Wharf
Beech Plantation
Low Shed
Coronation Belt
Hythe
Red Barn Pl
TROWSE WITH NEWTON

Subject	**Population Data, by Enumeration District**
Scale	Map (a) Population over 65 years - 2½" = 1 mile (1 : 25,000) Map (b) Population under 5 years - 1" = 1 mile (1 : 63,360)
Data	The data are drawn from the 1966 Sample Census of the General Register Office who supplied the enumeration district data and the boundaries. These maps handle only two of the types of data from the Population Census.
Digitising	The enumeration district boundaries were digitised at the scale of 6" = 1 mile. Time taken was 2½ hours for one person operating the digitising equipment.
Correction	The line segments were checked in the same way as the land use material (see page 40).
Output	In map (a), ED's where more than 18% of the total population are over 65 years old are marked. Map (b) shows similarly ED's with more than 8% of their total population under 5 years old. The boundaries of the relevant ED's and the symbols were scribed by the geagraph - the latter emphasize the areas meeting the specified criterion. Drawing time: Map (a) 11 minutes, Map (b) 6 minutes. Note: In the 6 square kilometres, no complete enumeration district is present - the calculations necessary to produce these maps were performed on figures for complete ED's.

a

Cottage Farm
Allot Gdns
Pit
Sch
The Chalet
Gravel Pit
White Farm
Mousehold House
Holmwood Wood
The Oaks
Weston Wood
High House
Thorpe next Norwich
Thorpe Lodge
Sch
Whitlingham Station
NEW CUT
The Carr
Rough Marsh
Plantation
Whitlingham Lane
Trowse Eye
Trowse Newton Hall
(Remains of)
White House
Old Wood
New Plantation
Low Shed
Oak
TROWSE WITH NEWTON

b

The Chalet
New Cut
TROWSE
WITH
NEWTON

Geology

These two maps were prepared as a part of another experiment for the Institute of Geological Sciences. The boundaries were digitised from field sheets at the 6″ to 1 mile scale. They were then played back on the geagraph at 6″ and 2½″ to 1 mile scales. At the smaller scale (page 68) masks for separate colours were scribed on the geagraph (approximately 2 minutes per colour).

Note that key numbers etc. were added by hand and not drawn by the computer.

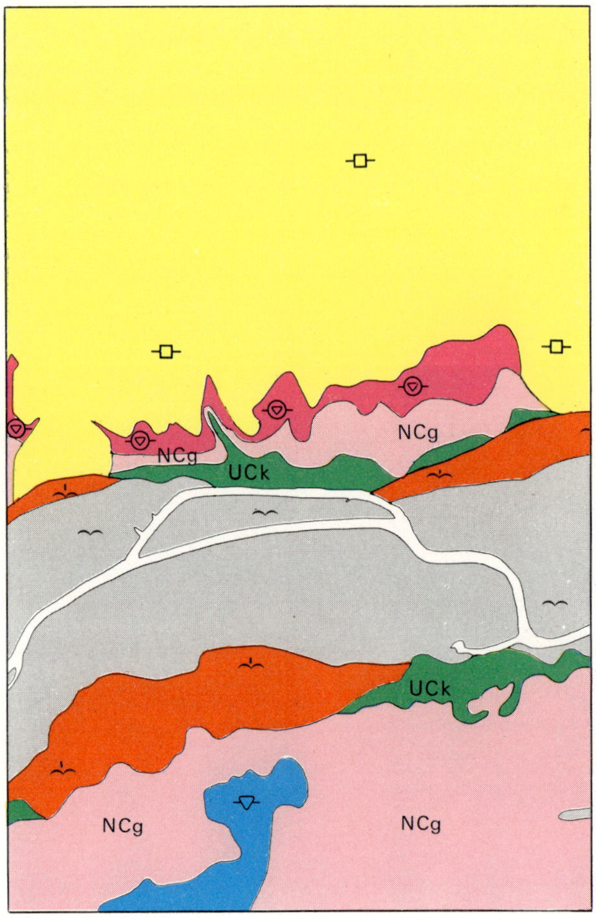

Legend

- Glacial Sand and Gravel
- Boulder Clay
- Brickearth
- NCg Norwich Crag
- UCk Upper Chalk
- Alluvium
- Terrace Gravels

THORPE ST ANN
TROWSE WITH NEWTON
Valley Drive
Cottage Farm
Allotment Gardens
Stone
Allotment Gardens
Club
School
Church
The Cottage (PH)
Brick Kiln Farm
Laundry Lane
Ring Road
Laundry Lane
The Chalet
High House Farm
Recrea
Gravel Pit
Church (Site of)
Garg Plant
White Farm
Athletic Ground
Morrow Pilling Park
Tennis Ground
Roman & Anglo-Saxon Remains found AD 1863
Pavilion
Gravel Pit
Weston Wood
High House
Allot. Gdns.
Playing Field
School
Holmwood
The Oaks
Pine Bank
Lion Wood
Newlands
Bracken Brae
Thorpe Tower
Western Ave
Stanmory Road
Wood End
Sunny Hill
Thorpe Lodge
St Andrew's Church
Parish Hall
Old Rectory
Hall
NCg
UCk
Road
Thorpe Old Hall
Boat Houses
Boat House
War Memorial
Boat Houses
Norwich 2
Yarmouth 20
Whittingham Station
Boat House
River Yare
New Cut
R D Bdy
b
b { Blofield and Flegg R D (Det No 2)
Thorpe St Andrew Ph (Det No 2)
Allot Gdns
The Carr
Rough Marsh Plantation
Whitlingham Lane
Post
Boat Ho Wharf
Trowse Newton Hall (Remains of)
Trowse Eye
White House
Old Chalkpi
Garden Cottages
Old Wood
New Plantation
Elm Cottages
Willow Cottages
Beech Plantation
Low Shed
Coronation Belt
Hythe
Oak
Red Barn P

Air Photograph Mosaic

The latest air photo cover for the area was ordered in the normal way from the Ministry of Housing and Local Government. Although the 12 or so photographs were not verticals of full survey standard, it was possible to prepare a mosaic from them: this has been photographed approximately to the scale of 6″ to 1 mile. (Note that there are minor distortions of positional accuracy.) The information on rateable value totals for grid squares shown on page 51 was then overprinted.

Section 5

Techniques of automatic cartography at the ECU

Section 5
Techniques of Automatic Cartography at the ECU

Digitising, computer processing and graphic output

Automatic cartography involves digitising, computer processing and graphic output. The first map proper to be produced by automation (crude lineprinter or tabulator-produced maps had been published earlier) was probably a bathymetric chart of the northwest Indian Ocean prepared by the ECU and published in October 1968 by the Royal Geographical Society. The isobaths on this chart, digitised on the d-mac table from 1 : 1,000,000 compilations by the National Institute of Oceanography, were scribed by the Geagraph with a line thickness of 0.1 mm. and at a scale of 1 : 2,000,000. This map demonstrated the ability of the basic system to produce accurate linework at least equal in quality to good manual scribing.

The digitising of map information converts analogue images into digital terms — x and y co-ordinates. The ECU currently uses a table with a manually-operated, free-moving pencil — a d-mac Mark 1B (CF1) Pencil Follower — with an active table area 1 m. square; this will be replaced (October 1970) by two new d-mac higher accuracy Pencil Followers, with active areas 60 cms. square. A stylus ('pencil') is used to follow lines, and it leaves a trace on lightly-coated material (masonscribe), that records which lines have been followed. A coil centred on the point of the stylus generates a magnetic field which is picked up by receiv-coils on a follower under the table. The follower moves in the 'y' direction, up and down a gantry which moves in 'x'. Taut wires transmit the 'x' and 'y' positions of the follower to shaft encoders which digitise them with a resolution of 0.1 mm. (accuracy 0.3 mm.). While we hope that table accuracy will be improved in the new equipment, the present accuracy is adequate for many purposes. Accuracy can be incresed by digitising at a larger scale; a scale reduction of 2 to 3 times produces accuracy up to Ordnance Survey standards.

Headings and labels can be input from a typewriter. Output is to the typewriter, or to an IBM 026 card punch, or (more usually) to magnetic tape. The IBM-compatible ½ inch 7-track tape is packed at 200 characters per inch — these are BCD characters in IBM 1401 code. Different labels are automatically attached to individual points, to points on continuous lines, and to corner points; end-line, end-loop, end-map, and characters from the typewriter are also distinguished.

Experience with a lock-on line-following digitiser (the 'geameter') revealed a number of disadvantages, in addition to relatively high cost and low speed. Until radically improved automatic digitisers are developed, manually operated equipment seems to be the most suitable — it is faster and cheaper, and it permits the use of rough or complicated field sheets.

A digitising equipment of the pencil follower kind costs some £10 k, and there are advantages in its being off-line to a computer. In contrast the drum scanning methods of digitising (see Section 6) demand very powerful and expensive computer support and seem to offer no very obvious compensating advantages for handling complex field sheets.

Processing is carried out on the Unit's Digital Equipment Corporation PDP – 9 computer. Core storage contains 16 k 18 – bit words, accessed with a cycle time of 1 micro-second. The hardware arithmetic unit performs integer multiplication or division in 19 micro-seconds. It is interfaced to the geagraph plotter, to an ASR teletype, two IBM-compatible 7-track magnetic tape units (working at 200, 556 or 800 characters/inch, with a speed of 45 inches/sec.), three DEC magnetic tape units (tape capacity 150,000 words; it takes 53 micro-seconds to read one 256-character block), a high-speed 8-track paper tape reader 300 characters/sec.), and a high-speed 8-track paper tape punch (50 characters/sec.). Total cost of the computer hardware is of the order of £50 k. A 262 k – word disc and a Tectronix storage cathode ray tube display have subsequently been added.

A limited amount of processing has been performed on larger computers but, apart from problems of compatibility, it has tended to be quicker to develop programs on the ECUs own small on-line computer. However, as data increase, and the requirements for a properly organised cartographic data bank emerge, so a more powerful computer – and one with a large disc store – becomes increasingly desirable.

The great bulk of plotting is done on the Unit's high accuracy drawing machine, the AEG Geagraph, which incorporates a 1.4 m. square Aristomat flat bed plotter. Resolution of the plotting system is 0.101 mm. and maximal plotting speed is 6.7 cm/sec.; however, the overall average speed (allowing for movement between lines, raising or lowering of the pen, and for slower movement on tight curves or through closely spaced points) seems to work out at 1 cm./sec. Plotting is carried out using either a 0.2 mm. rapidograph pen or a 0.1 mm. scriber.

Recent developments

Greater flexibility of output has become available since the completion of the project through this device, which is mounted on the Geagraph, projects a beam of light through:

a Light Spot Projector

This device, which is mounted on the Geagraph, projects a beam of light through a slot selected by program from a rotating disc containing 48 symbols. These can either be projected as point symbols, such as numerals, triangles and other more complicated characters, or as slits, which are moved over the film so as to draw lines. The system provides a tangential control, which ensured that the orientation of the slit is always normal to the direction of movement, a necessity if the line width is not to vary on corners. On the first operational disc (DISC 2), seven different widths of single line and four widths of double line are provided. Blank spaces are also available so that new symbols can be inserted ad hoc by sticking stripper film on to the disc in register. Any

Light Spot Projector disc showing symbols that can be called into position by program and projected onto film. Some of these are designed so as to be moved over the film to draw lines. Blanks are left for the mounting of extra symbols, ad hoc.

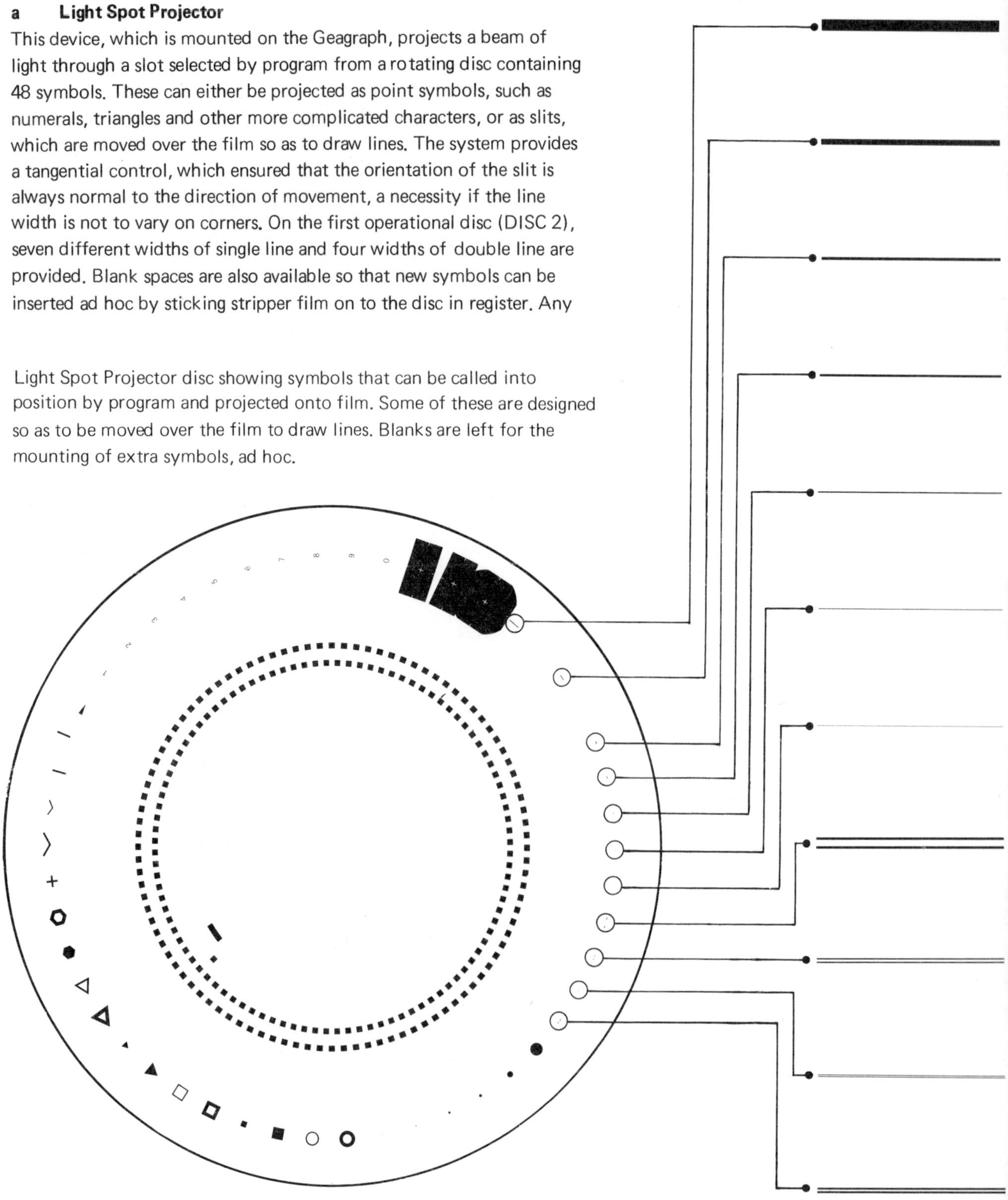

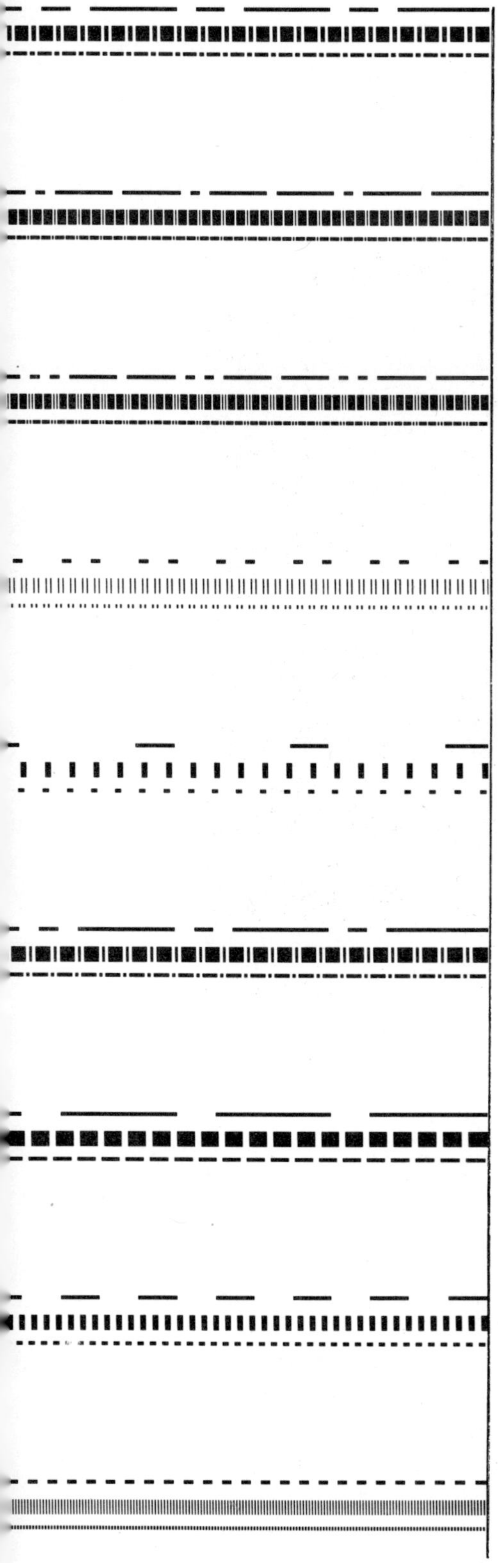

of the lines may be combined with any one of the nine possible pecking patterns which are selected by program but are generated by hardware. These pecked lines are formed from 'cells' made up by having the light on or off for any of 16 basic units (one 'cell'), the length of each unit being variable between approximately 0.2 and 1.3 mm. These lines or symbols are exposed directly on to stable base film, producing high quality photographic positives ready for lithographic printing. Examples of selected items from the range of possible output are shown on these pages.

Left: many variations of pecked line are possible by switching the light source on and off. Nine sets of pecking patterns are shown here. In each set the top two items show the same pattern on different line widths. The effect of stretching the pattern is shown by the third items.

Below: example of Light Spot Projector output: map at 2½" scale showing roads and railway in Bideford area

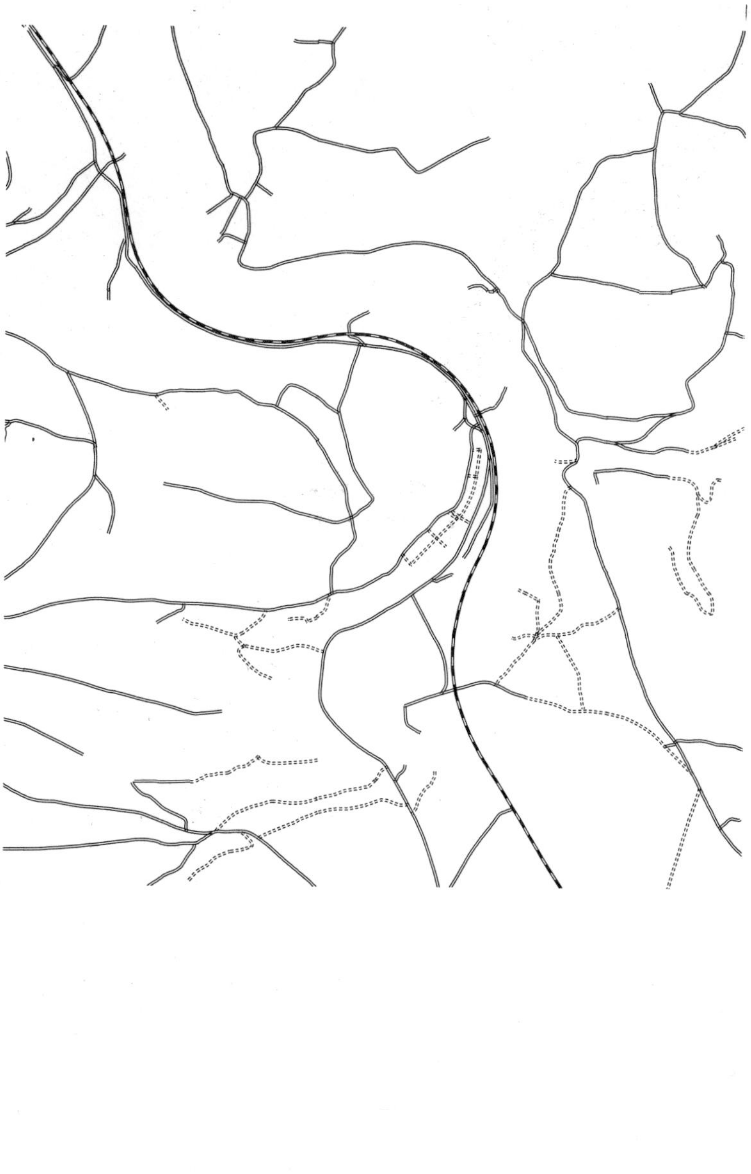

b Storage CRT

For many purposes a yet more rapid form of display is necessary, and is now provided by the Unit's Tektronix Storage Cathode Ray Tube display. Any shape of line which can be plotted on the geagraph can also be shown on this 16 x 16 cm. CRT in simple form and in about a tenth of the time.

CRT map of contours and water features for a 5 km. square in the Bideford area.
Plotted at 16 cm. square original size in approx. 20 minutes, then photographed from the CRT using a Polaroid camera.

c Photon 540

While well suited for its main purpose of drawing lines, delineating areas of colour, and placing irregularly distributed symbols at high accuracy, the Geagraph and Light Spot Projector combination is relatively slow where rough choropleth-style mapping of data collected on a grid square basis is attempted. Several attempts — e.g. SYMAP — have been made to design systems for the rapid production of maps from such data, and most of them make use of the standard computer output device, the lineprinter. This has certain obvious defects as the examples in Section 4 indicate. Recently the Unit has developed a method of this kind using a computer typesetter at the London College. of Printing — the Photon 540. Samples of this output from the pre-test

Scale 1:95,240. Size of symbol fixed (1.05 mm.) Type of symbol used indicates number of households per mapping unit (100 m. square). 4 classes used, indicated by dotless i, and three weights of 0

data for the 1971 census for the Huddersfield area are shown here, based on data made available by the Census Department of the General Register Office. Some typesetters are virtually as rapid as lineprinters; the Photon 540 is not in this class, producing up to ten symbols per second; however it has the advantage of being able to produce a vast range (1,400 characters) of different type faces, sizes and weights at any one pass. Furthermore it can accommodate a square cell size and produces output direct on film.

Scale 1:28,570. Size of symbol indicates number of households per mapping unit (100 m. square). 10 classes used.

Software

Though the existence of appropriate hardware is a 'sine qua non' of automated cartography, production of software is equally important. Most of the Unit's programs are written in Fortran 4, though some frequently-used routines are in machine code. Ideally software should be independent of hardware, but in fact considerable re-writing is often necessary to transfer programs to other hardware, because e.g. of varying formats and systems requirements. The ECU, however, has a program which writes data tapes in a standard format for export to non-DEC computers and the equivalent capability to read data tapes from other organisations if they are in this format. Some notes about certain basic programs are given below.

Feature coding

Each kind of line or point is differentiated by a five-digit feature code which is typed in at the time of digitising. If no new code is given, a new line automatically inherits the feature code of the previous line. Since code-typing is a distracting and error-prone operation, there are advantages in digitising all those lines which have the same codes in one shot.

The plotting of a new map involves specifying the feature codes required. Multiple coding is necessary both for relating statistical values to point locations, and also for labelling boundaries according to the areas on either side, e.g. formations on geology maps or different classes on land use maps. In this way it is possible to draw, for example, only those lines which bound residential land, and this drawing can then be used for the production of a colour mask (see Section 4). Multiple coding is also necessary for handling an administrative boundary, which for example might need to be coded for two county names, two rural district and two parish names.

Co-ordinate transformation

Another capability of fundamental importance is to change from the arbitrary co-ordinates of the digitising table, to an absolute co-ordinate system. Before the digitising of any sheet commences, the corner points are digitised. At the computer processing stage, the desired co-ordinates of these corner points in a stated absolute system are typed in. The program then makes a least-squares fit of the first set to the second, to establish the correspondence between co-ordinate systems. The co-ordinates of all detail on the map are transformed to the absolute system by applying the same bi-linear transformation required for the corner point positions.

This program permits maps digitised as several separate sheets to be reassembled automatically, with the sheets in their correct relative positions. It does not ensure that lines cut by a sheet edge meet exactly, for if the sheets have been digitised independently there may be operator or digitiser error in opposite directions on the two sides. Hence, it is essential to check correspondence along sheet edges, before digitising commences; this is especially so for lines which have been sketched in the field, as in soil and geology surveys.

Co-ordinate transformation can rectify linear distortion in the original document: this is especially useful where there is no alternative to digitising from paper field sheets.

Tape correction and up-dating

It is, of course, crucial to be able to check and correct errors during production, e.g. after the material has been digitised; and exactly the same facilities are required to up-date tapes later on. Correction involves the process of deletion and of addition. The requirement to edit tapes for these two purposes dictates many features of their format, and dominates the design of an automated cartography system. It would be very irritating to have to discard a whole tape because a few crucial errors could not be corrected.

The checking of digitised data involves two aspects –
checking for positional accuracy
checking of feature coding

The check for positional accuracy involves playing the data back on a transparent overlay on the high accuracy plotting table and to the exact scale of the original. Where very few feature codes are involved, it is sometimes possible to combine the positional accuracy check with the feature code check by playing back different codes in different graphical forms – i.e. with different colour rapidograph pens or different symbols and pecking patterns on the light spot projector. However, where for example one line has several different codes it is necessary to draw them several times over.

Where many different feature codes are involved it is convenient to play the positional check in a single symbol, and to check the codes separately. The codes are checked by producing a printed list specifying for each item its co-ordinate (or start and end co-ordinates for lines) and its list of feature codes. Each item is identified by checking the co-ordinates and the feature codes can then be verified against the original document from which the digitisation was made.

Corrections are applied to the data as stored on magnetic tape by means of a correction program. This program permits the user to produce on the digitiser a special correction tape. The correction tape may specify such commands as "delete the line which passes through the following digitised point", "change the feature codes of the line which passes through the following digitised point", or "add the following digitised line". The correction tape is merged with the old data tape to produce an up-dated data tape.

An alternative technique for feature codes is to employ a program which allows the user to check and correct inter-actively. This program draws the data on the CRT storage display one item at a time. After each item is drawn, its feature codes are typed out. The user is then invited to (a) accept the codes as they stand, (b) make corrections, or (c) delete the item altogether. This inter-active approach while obviously more expensive in computer time, obviously reduces the overall time to complete the checking task.

Data compression and generalisation

The fact that ten BCD characters are required to store the co-ordinates of a point digitised on the d-mac, makes the use of magnetic tape essential for the recording of lines. The spacing of points along lines depends on the speed with which the follower is moved, and this depends in part upon the intricacy of the line. Usually there are from 1 to 3 points per mm: there are more points on sharp bends and fewer on straight lines (in fact if a line is known to be straight the operator need only record its two ends). Hence, the digitisation process itself performs a useful data compression, in relation to the total number of points which could be digitised.

The use of cubic splines for data compression does not compress data sufficiently to compensate for the computing time and later decoding involved. However, to plot maps at greatly reduced scales it is necessary to discard many points, i.e. to generalise shapes, both for graphic appearance and to maintain a reasonable drawing speed. The original software attempted this by omitting intermediate points provided that they lay within a specified distance of the line drawn. This was unsatisfactory because the computing time involved, especially for extreme generalisation, took so long on the PDP-9 that it was slower than drawing every point.

Following experiments with different techniques (Lang, 1969) it was found that satisfactory results were obtained by either: (a) retaining every nth. point; or (b) retaining only those points more than a certain distance from the previous retained point. Attempted speed is inversely proportional to the distance between selected points, and the minimal speed is set at 13.3 mm./sec., i.e. 20% of the maximum. Actual speeds are somewhat lower, because of acceleration and deceleration for changes of direction. A point spacing of about 0.3 mm. at the scale of plotting produces a map which is neither too detailed, nor too angular. The time taken is at least 30% less than with the original software, and more time may be saved by greater generalisation. There is, however, plenty of further room for experiment with automatic generalisation of shapes.

Anaglyph maps

A field in which automated cartography has great advantages over conventional methods is the production of anaglyphs or stereo pairs to represent the third dimension. Topographic surfaces, geologic boundaries, air or water mass boundaries, or more abstract surfaces, may be represented in this way. Given digitised contours labelled with the appropriate heights (or 'z - values') the program is relatively simple. For each half of the pair, each contour is displaced in proportion to its height. The left and right displacements are set side by side for a stereo pair, or superimposed in red and green respectively for an anaglyph, which must be viewed through spectacles with tints exactly matching those in which the map is drawn.

It is necessary to specify the proportionality constant between height and displacement, and the level of zero displacement. The proportionality constant must be chosen carefully to produce an appropriate relief at a given viewing distance (Adams, 1969). Relief should not exceed 3 to 5 cm. over a short distance (the field of view), for a viewing distance of 35 cm. With automatic plotting, a good three-dimensional effect can be achieved, whereas manual redrafting of contours produces differences which make it more difficult for the eye to fuse right and left images. If a separate plate of each contour were duplicated photographically, there would be difficulty in combining them with exactly the right displacement. Hence, it might almost be said that automation makes the large-scale production of anaglyphs and stereo pairs from contours feasible for the first time. An example of an anaglyph map is shown on page 61 in Section 4.

Contouring

The ECU has a number of different contouring programs. The simplest of these can accept up to 3,000 irregularly distributed data points while a more sophisticated scheme which uses the surface slope at each data point as well as its height can accept up to 1,350 data points. These would normally be digitised on the d-mac, but points digitised manually could be input when converted into the appropriate format. In the simplest case, height values at points on a regular grid (120 x 120 or smaller) are interpolated by applying a weighting function to the known values at the nearest 4 to 10 data points. Data points within a radius likely to include 7 points on average are examined: if there are more than 10 such points, the nearest 10 are used: if there are less than 4, the radius is repeatedly increased by 2 until 4 points are encountered. These points are then averaged after being weighted by the function $(1 - R^2) / R^2$, where R = (distance from the grid node in question to the data point) / (radius of area examined). R varies between 0 and 1. The use of this weighting function produces a continuous surface without necessitating examination of all points for each grid point.

In a second phase, the resulting height grid is scanned once for each altitude for which a contour is to be drawn. When two adjacent grid values bracket the desired altitude, the position of the contour between the two points is estimated by linear interpolation, and stored. When the scan is complete, these interpolated contour positions are arranged into the drawing sequence, and joined by smoothly curving cubic functions in x and y which are plotted by the geagraph as a series of short straight chords.

The resulting map is much better for representing smooth surfaces than for rough surfaces where the data points are distributed irregularly; however, such data cannot be contoured with confidence manually. Further variants of the program, introducing smoothing effects, are used to improve the appearance of maps produced from poor data. The program is being transferred to a larger computer with much faster computation, and it will then be able to handle more points and work through larger grids.

Length and area measurement

A simple program to calculate the lengths of lines and the number of points on them has provided very useful measurements of the spacing of points in large sets of digitised data. The lengths of lines with different feature codes are cumulated separately.

The program for measuring areas performs a Simpson's Rule integration for unequal intervals. It is not necessary to use the digitising table simply as a planimeter: if several lines bounding an area are digitised separately in the normal way, the areas between each line and the x - axis can be calculated by program, and totalled outside the computer. Due regard must then be paid to sign; area is given as positive for a line digitised in the direction of increasing x, negative for decreasing x. We hope to be able eventually to tie such segments together automatically.

Section 6

Other research and development in this field

Section 6
Other research and development in this field

The Canada Geographic Information System (Canada Land Inventory)

The Canada Geographic Information System of the Department of Regional Economic Expansion of the Canadian Government is being developed jointly with IBM. Work started in 1965. Not all of the system has yet been accepted by the Department.

The system is intended to accept, store and process information relating to agriculture, forestry, present land use, administration, census, watersheds, recreation, and wildlife (waterfowl and ungulates). It is designed to run on an IBM 360/65 with 517 k bytes core store, 6 tape drives and 4 disc drives. It can also be used on a smaller system (IBM 360/50); currently it is run on an IBM 360/85.

Input to the system is taken from maps of single variables – e.g. classification of land for forestry. The digitisation of each map requires the preparation of 3 distinct sets of data :-

(i) Geographical boundaries. The map describing the boundaries of interest is transferred photographically to an opaque plastic material. The boundary lines are scribed out manually, leaving transparent line-work. The resultant map sheet is scanned on a rotary "drum scanner". This scans the map as a matrix of small squares (of side 0.004 inch) and produces on magnetic tape a sequence of "1" or "0" bits according to whether a line is or is not present at the corresponding position on the map. The scanning process takes about 15 minutes for a 48 - inch square map, and proportionately less time for smaller maps.

(ii) Area identifiers. The individual parcels ("faces") are given sequential identifying numbers, which are marked on an overlay. The visually estimated centres of each area(any internal point will do) are digitised on a d-mac table strictly in the order of the identifying numbers. The digitised co-ordinates are written on to magnetic tape.

(iii) Area classification data. A list of the numeric data describing each face is prepared and punched on cards.

Several computer programs are then run to amalgamate the three sets of data for input to the data bank:-

(i) The face identifier co-ordinates and face classification data are merged into a single tape.

(ii) The "bit-matrix" produced by the drum scanner is processed to separate out the individual line segments as strings of x, y co-ordinates.

(iii) The individual line segments are then associated with the two faces on either side, using the locations of the visual centres of the faces to provide the necessary linking information.

The input process is the most demanding stage of the system, involving large programs, of which many parts are written in assembly code. At present most maps are digitised at a scale of 1 : 50,000, and only 80% of map input can be processed correctly. A typical map has between 500 and 3,000 faces.

Within any one map series the information is grouped by rectangular geographical areas of "frames", each 1/6 of a 1 : 50,000 map sheet. Co-ordinates are stored in compacted geographicals to an accuracy corresponding to one-fifth of the resolution of the scanning process (i.e. 0.02 inch on 1 : 50,000 maps). These rather coarse steps are smoothed at the output stage.

The data for each frame are further subdivided into the image data set of co-ordinates specifying the face boundaries, and the descriptor data set containing the descriptive information for each face (the classification data supplied on input, the face area, and centroid co-ordinates). This facilitates processing of large descriptor data sets. The data sets are cross-referenced inside the computer by a system of pointers.

As yet the general system for updating the data bank has not yet been accepted from IBM. Corrections are made by re-digitising a whole map and adding the corrected data to the end of the existing data as if it were a new map.

For a data bank enquiry, an area of interest has first to be specified. The alternative forms of specification are:-

(i) topographical sheet number,

(ii) an area already digitised and stored within the data bank, i.e. an administrative unit, a census unit, or a regional development area,

(iii) a circle of specified centre and radius,

(iv) a polygon of specified corner points.

Queries may be directed either towards a single subject material, e.g. "find and plot all the class 3T agricultural areas", or to multiple subjects, e.g. "How much class 1 to class 3 recreational land is currently unproductive, or used for pasture, grazing, or woodland?" The inter-relationship (i.e. "overlaying") of different subjects is achieved by converting the spatial distributions into a grid or matrix form and then relating the values at corresponding matrix nodes so that an 'overlap' map can be produced.

Programs to answer standard queries will be built into the system. Non-standard queries will require new programs, written in PL/1.

The data bank interrogation programs have been run for demonstration purposes. The two main disadvantages of the system are the expense of digitising, which, because of the need to redraft, is in effect done both manually and automatically; and the computing intricacy of sorting out drum scanner records. However, a major software effort has been undertaken, particularly with techniques of data storage, indexing and compaction, and aspects of this may be of more general application. The ability to overlay up to eight features in the computer and produce maps of combinations is particularly valuable, as will be the ability to search and retrieve data from a large data bank. Processing costs are currently some £200 per map, with computer processing time being the major expense.

Line Printer Mapping

Both the ECU and CGIS systems are specially designed for computer-controlled cartography. Another approach has been to use more widely available computer peripherals to produce low-precision cartographic output. Drum plotters and line printers are the two peripherals most commonly used and cathode ray tubes are also coming into use.

All major computer installations have 'line printers' for the rapid output of 'hard copy'. The symbols used include digits and letters (often upper case only), as well as asterisks, brackets and other symbols commonly found on typewriters. Most line printers have characters 1/6 x 1/10 inches (4.2 x 2.5 mm.), and print 120 or 133 of these to a line. Since a line of symbols is printed simultaneously, it is easy to shade in a whole area to produce a layer map or a choropleth map. Line printer output in general is on unstable paper, and it is difficult to superimpose on a base map; no line printer can produce acceptable lines.

Numerous line printer mapping systems have been produced, a reflection of the simple programming involved and the ready availability of the required equipment. The most important of these will now be evaluated.

Meteorological Office

An early use of line printers for mapping was in the Meteorological Office, which produces hundreds of isarithm weather maps each day by line printer. Alternative inter-isarithm bands are identified by digits, and intervening bands are left white, giving 20 levels; letters and other symbols may be used to give more levels. The isarithms (of temperature, or pressure) defined by the margins of these bands are traced manually onto an overlay if the map is to be transmitted via a scanner, since firm lines are needed to provide a clear signal. To cut out this redrafting stage, the Meteorological Office is now using an

Electronics Associates Vari-plotter for very high-speed, low-accuracy (0.015 inch, 0.4 mm.) isarithm plotting, as an alternative to and perhaps eventually a replacement for the line printer. This sets the line-printer in perspective as a 'stop-gap' mode of isarithm output, for use in the absence of a plotter. In either case, the low-accuracy of output is of no concern given the coarseness of the data, and the fact that many maps are based on extrapolation into the near future, so that the uncertainty is much greater than the cartographic inaccuracy.

Similar isarithm output on line printers is often produced by trend surface programs, such as that of O'Leary et al. (1966). This is, however, a rather specialised application, since the surface level at each printer position is evaluated not by interpolation, but from the equation defining the trend surface.

LINMAP
A range of line printer mapping and map processing programs developed by the Urban Planning Directorate of MHLG has been described by Gaits (1969). Since this project has already produced a large number of maps, it merits attention separate from other work by planners.

LINMAP system programs are written in Fortran 4 for the CDC 3300 computer. Data are prepared by digitising data points and corner points on a d-mac. The National Grid co-ordinates of the corners are recorded on the tape after the four corner points have been digitised. The tape is divided into groups for which a local authority code is given, followed by the number of points (i.e. zones) in the group. The body of information on the tape consists of triplets representing the x and y co-ordinates of a point (representing a zone, i.e. a ward or parish) and a reference identifying the zone within the group.

This tape is checked and edited (via the paper tape), then table co-ordinates are transformed into National Grid co-ordinates. After a further check, this location reference tape is merged with GRO data tapes which use the same local authority and zone coding. A seperate file is produced for the boundary of the area to be mapped; this is limited to 3,000 points representing vertices of the boundary. Up to 5 maps, each based on any combination of up to 4 'z' values from the GRO information, can be produced in one run.

LINMAP uses line printer symbols to represent either absolute quantities, or percentages, ratios etc. The DOTMAP option allots parish or ward data to the printing position in which the centroid lies; this leads to amalgamation of closely spaced units and — more unfortunately — gaps for positions which, though lacking a centroid, do contribute to the data. Such gaps are in order when absolute values are mapped — so long as the range of point symbols is well designed — but they are misleading for percentage maps. Hence LINMAP 2 contains the SURFMAP option, which produces an approximation to an isopleth surface by interpolating between the filled printing positions. Other additions include choropleth mapping for grid squares (GRIDMAP) and for irregular areas (ZONEMAP), and a coloured version of DOTMAP output (COLMAP).

The primary advantage of the LINMAP system is that it has provided rapid cartographic access to the whole 1966 census file. Points for all wards and parishes in the country have been digitised, converted into National Grid co-ordinates, and merged with GRO magnetic tapes containing census statistics. The ability to produce maps of either original values or ratios, at any scale, is a major achievement. The next stage is to improve the cartographic representation. Like SYMAP, LINMAP appears to have an inversion in density in part of its ten-level 'grey scale'.

SYMAP

The most widely used computer mapping system is SYMAP (Robertson, 1967; Rosing 1969), developed at Northwestern University and the Harvard University Laboratory for Computer Graphics under the direction of H. T. Fisher. It was featured at a recent PATRAC conference, the proceedings of which will be published in 1970. SYMAP is designed to use a standard line printer, but because all its options are combined into one package it requires a very large core store, and can run only on large machines. Various people have attempted to reduce its storage requirements by simplification or division into smaller packages with fewer options, while others have decided to start from scratch with new systems such as LINMAP.

SYMAP is known to be running at the following centres in Britain:

Location	**Machine**
Durham (University Computing Centre)	IBM 360/50 (line link to Newcastle
East Kilbride (National Engineering Lab.)	Univac 1108
Edinburgh (Regional Computing Centre)	IBM 360/50
London (University College)	IBM 360/65
Manchester (National Computing Centre)	ICL 1905 F
Newcastle (University Computing Centre)	IBM 360/65

The NCC has undertaken to distribute SYMAP in the UK at £250 per set of cards and literature. SYMAP produces contour, proximal or choropleth maps. The production of choropleth maps is rather inefficient in that the outlines of each parcel must be digitised, i.e. all vertices must be digitised twice. Since boundary co-ordinates are rounded to the nearest printer position there are usually a number of overlaps or white spaces between parcels, where slight differences in the co-ordinates digitised have been rounded in opposite senses. The difficulties of data preparation make the program suitable only for the production of large sets of choropleth maps for similar geographic units; this, however, is often required from census data for administrative areas. The program is not very efficient in terms of computer time.

Proximal maps are less demanding. They take the same point data (x, y, z,) input as contour maps but instead of interpolating values for intermediate positions, each printer position is given the 'z' value of the nearest data point. This produces a map of irregular polygons, approximating to first-order Brillouin regions. Because each printer position must be calculated, proximal maps take as long as contour maps, despite the simpler computation.

SYMAP is most commonly used for contouring point data by interpolation which is primarily distance-weighted, but also takes account of 'shadowing' and local gradients. Since all points are stored in core, this limits program capacity to 1,000 points in the standard implementation of the program. The resulting surface is generally satisfactory, and passes through all the data points. 'Barriers' across which the weighting for interpolation will be reduced can be input, but their use is not recommended since they increase computing time about twenty-fold. Also, the 'barrier' routine weights the interpolation by the 'permeability' of the barrier, but still uses the straight-line distance between data point and interpolated point, whereas in many situations the replacement of this distance with the distance around the end of the barrier (e.g. the lowest bridgepoint on an estuary) would be more appropriate.

Another facility of limited use is the printing of legends, since the output is a poor return for the considerable effort of input. Point and line symbols, however, are of value in improving interrelation with a base map. The option of producing histograms and summary statistics is of considerable value, but the output format is wasteful. More compact tabulations and 'horizontal' histograms should be used, since at present the maps tend to be buried in a mass of often unused output. Repeated print-out of input for each map in a

Output from SYMAP is usually in terms of a grey scale of up to 10 levels: larger numbers of levels could be produced by amalgamating separate runs of the program, but this would involve duplication of symbolisation so that redrawing might be necessary (e.g. to produce more than 10 contours). The fact that shading and not just contours are produced is howevery very useful, and permits more immediate appreciation of the surfaces portrayed. The shading is produced by different printer symbols, overprinted at the end of the scale. Unfortunately there are inversions in the full, standard grey scale; two levels appear lighter than lower levels, contrary to the general black-to-white downward progression. This disadvantage can be overcome by a user specifying his own set of symbols, but the fact remains that design of an appropriate set is difficult within the constraints of the standard line printer symbol range. A user with a dedicated line printer could design a new range especially for the production of grey scales, and this would also eliminate the need for overprinting; but this would only be worthwhile for a very large output of maps. Currently, the best visual results are obtained by not using more than six levels.

Contour 'lines' can be produced either as white edges to the grey-scale layers, or in black without a grey scale. The resulting low-resolution lines are visually unattractive, and quite chaotic if contours are closely spaced. Hence only use of the simple grey-scale layer map is to be recommended.

Map size is not limited by line printer widths, for maps can be produced in many separate sheets. This permits the production of large maps for photo-reduction, so that the coarse (4.2 x 2.5 mm.) symbol size is reduced to give a better visual impression of greyness. There are matching problems when one side of the line printer prints more darkly than the other (e.g. through the greater use of the left-hand columns). K. Rosing has overcome this problem by using special three-layer paper, producing an even image by use of carbon paper.

This summary has emphasised the limitations of SYMAP, but its wide use indicates its value. It would be useful to have its three sections running as separate programs, suitable for smaller computers. Even with large computers, charges are kept down by reducing core storage requirements. Separate choropleth, proximal and contour maps (without barriers) are needed, and the standard symbolisation should be changed. The speed of line printer mapping is very useful, no special plotters are required, and such programs might be of great value where accurate maps were not required.

The University of Edinburgh
Several mapping programs have been developed in the University Geography Department. One of these is CAMAP, a choropleth mapping program written in Atlas Autocode but similar in some respects to the relevant part of SYMAP, except that it is very much faster and demands less core store; input is from magnetic tape.. CAMAP utilises overprinting with symbols on a line printer. Its principal use to date has been in producing parish-based maps at 1 : 2,500,000 scale for the Agricultural Atlas of Scotland. Photographic reduction of the maps for publication produces pleasing results, illustrated in Coppock (1969). Colour maps have been produced by making plates from separate maps of the distribution of each level, e.g. 50 x percentage cover of crop in the parish ix 75. Line printer-produced block diagrams based on height values on a rectangular grid have also been produced.

A much larger and more general-purpose mapping system has also been implemented on the IBM 360/50. This program was originally designed to handle census style data based on enumeration of similar districts. Input may be from any medium, and thereafter a series of options specify what is to be selected and how it is to be manipulated statistically and, finally, mapped. Either sections of the processed data, or the map, can be filed after the run for subsequent re-use with change of scale, etc. Output is generally by line printer but may alternatively be routed through a Ferranti Micro-plotter giving microfilm record inserted in punched cards.

Other Modes of Output

Drum plotters

Many computers have a fast, low-precision drum plotter, e.g. a Calcomp, for graphic output, and these have some application to planning. These include histograms, block diagrams, perspective drawings for planned highways (a service also available from Hunting Technical Services and Compugraphics), contouring, weather charts, subdivisions for housing developments, and mapping with point symbols. The ability to draw lines obviously gives much greater flexibility than is available with a line printer, while it is possible to draw digits, letters and point symbols by specifying a limited number of vertices. The main disadvantage of such plotters is their low precision (usually 0.01 inch, which produces visible steps on most lines) and the use of relatively unstable paper to draw on. The plot is usually in biro, and has to be redrawn for reproduction. However, for 'one-off' tests, drum plotters may be adequate, and their speed of operation make them relatively inexpensive. The character-drawing facility should not be over-used, since it is relatively slow (compared with a line printer). Some (more expensive) models incorporate a printing head as well as a pen, which improves speed of character plotting.

A Benson-Lehner plotter has been used by Hawkes et al. (1968) to represent botanical distributions in terms of frequency and environment for grid squares. Lines are 'thickened up' by drawing closely-spaced parallel lines. The use of a Variplotter by the Meteorological Office has already been mentioned; this is actually a flatbed plotter, but is akin to drum plotters in its low resolution and accuracy. Smooth lines through widely spaced points are produced essentially by the inertia of this high-speed device.

Cathode ray tubes

That cathode ray tubes have not yet been used widely for map production is due partly to the expense of the interactive type (and the size of computer needed to back them), and partly to the distortion inherent in most tubes. Both these situations are improving, and useful maps have been produced on the Ferranti Microplotter at Edinburgh; but the accuracy is still inadequate for superimposition on a topographic base map. Simple dot maps have been produced on the SC 4020 at Aldermaston, and the ECU have used this device to produce grey-scale output from a hill shading program, but the quality of image is disappointing. A sample of the output from the ECU's own storage Tektronix tube is illustrated in Section 5 of the report.

This is a field in which technology is advancing rapidly, and cathode ray tubes or their successors are likely to be important cartographic output devices.

Tabulators

Another output device is the 40– or 80– column tabulator. This was used to produce the grid-square-based dot maps in the Atlas of the British Flora (Perring & Walters, 1962), which was probably the first published set of maps produced through automation. Open or filled circles of several sizes are placed at positions arranged in a square grid, under the control of punched card input. In this case statistics are known for each square. A similar device has been used by Bertin (1967) particularly for the production of choropleth maps of France. Hence the statistics apply to each Département, and all grid squares falling principally within that unit are given a dot of approximate size. Though the resolution is very coarse, the visual effect is good because the unreal harsh, irregular boundaries of normal choropleth maps are suppressed.

Other computer processing of locational information by planners

This section attempts only to be a very brief summary of some of the major pieces of research that were encountered during the course of this project. It may provide an approximate guide to the state of development and application of computer maps for planning purposes in the UK.

One of the basic problems has been that of timing. The rate of progress in this field is extremely rapid and, in many cases, the work described will have been superseded by new developments. Conversely, there have been many occasions where possible future developments have been discussed and it is difficult to differentiate between long-term hopes and short-term achievements. Thus, some of the projects mentioned may by now be in operation, while others may have been abandoned.

Some of the work described is being undertaken in the planning departments of local authorities, some in Universities, some by government agencies and some by independent organisations. Much of the work, whilst dealing with planning data in relation to computers, does not as yet involve computer mapping, though there is obviously potential for mapping.

London Borough of Southwark

The planning department is operating a computer based land use record. Information relates to land parcels (which may be groups of properties) and all information includes codes for street block, land parcel, street address, enumeration district, traffic zone and OS 1 : 1250 map sheet. The street address code is formed from the house number (or numbers) and a street code according to a scheme for the entire borough. Alphabetic suffixes (e.g. for blocks of flats) are treated separately and are allotted working backwards from 999 within each street. An alpha-numeric system would be preferred but the arrival of a new computer is being awaited before the system is changed. Analysis is done on the LEO 3 machine at Greenwich which belongs to a consortium of SE London boroughs. The process of revision on the ground is a continuous one, taking two years to cover the borough. The information is put onto magnetic tape about once a month and the actual up-dating of the land use record is done every 4 - 6 months (incorporating 4 - 6 revisions at once). Replies to analysis requests take about 3 or 4 days but, in cases of extreme urgency, can be completed within 12 hours.

One problem that has been encountered is that of assigning the site area in mixed use land parcels. An analysis of total site areas in particular land uses results in double or triple counting of site areas where parcels are in mixed use: proportional splitting of site areas would often be unrealistic.

When the new computer is installed, it is planned that the system will be extended to cover information on planning applications and decisions, extending back to April 1965, when the GLC was formed. Hopefully this system will be cross-referenced to the land use classifications and also to the information which it is intended to include from employment exchanges.

About two years ago there were some experiments in mapping traffic zone data using a modified form of SYMAP. Problems arose in that the Greenwich machine would not over-print several characters on the same space, and so the work was done by a firm of consultants. In this case it proved to be quicker for the Southwark planners to map the material manually.

West Riding County Council
(and Yorkshire and Humberside Standing Conference). The Council has extensive plans for all-purpose data banks at property and enumeration district level. These would serve other Departments besides Planning, and might involve computer mapping. Work has started on allocating to 100-metre grid squares new data collected by the Medical Officer of Health. For the data banks, each data item will incorporate a wide range of identification and location codes, including national grid co-ordinate references, local authority code and enumeration district code. WRCC planning department have coded and analysed a large amount of data on the basis of grid squares. A variety of information (population, land use, employment, soil and foundation conditions, etc.) has been presented by 100-metre grid squares at 1 : 25,000 scale and by 1 km. grid squares at 1 : 63,360 scale. The latter is used for more general analysis at a sub-regional level and for presenting confidential information in aggregated form. For presenting land use, each 100-metre square has been assigned to a single land use category on the basis of the dominant use, and this produces distortions; an important feature such as a motorway cutting diagonally across the graticule shows up as a discontinuous feature, appearing as the dominant use in some squares, while absent in intervening squares.

Liverpool City Council
The planning department is collecting a large amount of very varied data, particularly in connection with a study of "social malaise". Many of the data are in machine-readable form, and the long term intention is that they should form a data bank. Street address and enumeration district codes are attached to all information where confidentiality permits. A grid reference may be added later. The street address code consists of the house number and the postal code (unique to each street). Ward figures are now complete for this study but no computer mapping is being undertaken at present.

Other surveys which are in progress are in connection with employment, re-location (all firms affected by motorways and housing proposals), pedestrians, planning applications and origin and destination in connection with re-housing. During November 1969, a detailed report of progress with the use of the computer in the planning department was being prepared.

East Sussex County Council
The planning department deals annually with about 10,000 applications for planning permission through the 15 different local authorities, and a system has been evolved for recording these applications on aperture cards (i.e. punched cards with a small piece of film inserted). A small projector throws the film (say, part of an OS map) onto a more or less horizontal surface; notes and additions can be made to this projected map and re-photographed to record permanently the amended version, whilst still retaining the original intact.

Hertfordshire County Council
(In collaboration with Essex County Council) A systems analyst has been employed to investigate methods of extending the use of computer techniques in the planning department. One study topic that may involve mapping by computer is the record of sites available for residential development, together with records of planning approvals and building completions. The objective is to show from time to time where land zoned for residential use is the subject of approvals for residential development, and where development has or has not taken place. This would be a complex operation as several planning approvals can relate to the same site, or over-lapping sites, and schemes may be implemented on a partial basis. A highly accurate large scale site record would be required, incorporating a flexible retrieval system so that numerous combinations of sites could be assembled.

No single system for locating data is envisaged but rather a combination of different codes, as considered relevant to different subjects, probably including a grid reference (up to 12 digits). Paper tape will be the input medium. The Council has not yet decided upon their choice of digitising machinery.

London Borough of Brent
In connection with building up an information system in the Borough, the planning department has developed a 20-digit referencing system for data items consisting of the 1961 Census District, enumeration district, 500-metre grid square, street block, land use parcel and street name and number. Next year

They hope to set up a central transcriber file with the full range of reference to each of the 100,000 properties in the borough, as used by each of the various departments.

Birmingham City Council

There are plans to develop a data bank for Birmingham, initially for the planning department but eventually extending to the other departments. Work has begun on digitising individual hereditaments, with a 12-figure National Grid reference. The external property boundaries are also digitised, giving the net and gross acreages (gross includes half the surrounding road width and is used for density calculations). Planning applications and searches have high priority for inclusion in the data bank, although the digitiser is also used for allotting GRO information to grid squares, and dealing with highway and service (e.g. gas) studies.

University of Essex (Professor Potter)

This is the home of the SSRC data bank which was set up in October 1967. Computerised survey material from numerous academic, governmental and commercial sources is referenced and stored, and is available for reprocessing by other users, subject to certain safeguards. Apart from a list of the surveys, the main index is the KWIC (key word in context) system based on the major words included in the questions/replies in the questionnaires. Where possible, questionnaires are also retained. Geographical resolution and coding of data varies from survey to survey, and is dependent on the needs of the initial sponsorts of each survey.

University of Reading (Geography Department)

Work has recently started in the Urban Systems Research Unit on information needs for planning. The problem is being approached from a systems point of view, and it is hoped to develop a comprehensive conceptual model of the urban system. This will indicate the types of data that are required. The work is in its early stages, and no computer mapping is involved at present.

General Register Office

The GRO has been considering the use of location codes in the 1971 population census. Pre-tests have shown that enumerators can allocate households to 100-metre squares where large scale maps exist; in other areas a kilometre grid may be used. If such a system is adopted, data for 100-metre grid squares will be released only where there are enough cases to safeguard confidentiality. The possibility of giving each household a unique co-ordinate reference using d-mac equipment is also under investigation.

Business Statistics Office

The Business Statistics Office (BSO) has been developed from the Board of Trade Census Office; to be staffed and run by the Board of Trade, it will have a small interdepartmental management committee under the Chairmanship of the Director of the Central Statistical Office, to direct policy. The BSO has the task of building and maintaining a Common Register of Businesses to be based on an ICL installation at Newport, Monmouthshire.

The core of the system is likely to be an index – carried on magnetic tape and in other forms – of the names and addresses of business establishments. Basic information will include name and address, activity, numbers employed, employment exchange area, turnover etc. This "list", classified in a number of ways, will provide a master frame for surveys and samples and will make for more efficiency in the collection and use of statistics. Initially the basic geographical reference will be the employment exchange, but serious consideration is being given to an individual co-ordinate referencing system.

Raw data about individual concerns will be highly confidential and, for most purposes, they will be aggregated to employment exchange level, which will provide a wide range of non-confidential data on a standard basis; re-processing of data tapes could take place to answer special queries when necessary.

West Midland Regional Study (Mr. Stevenson)

Computer mapping has taken place on two fronts using both LINMAP (from the MHLG) and SYMAP (from the NCC). The LINMAP work has involved the mapping of 1966 census data over the whole region on a ward and parish basis using a scale of 1 : 250,000. A number of maps has been produced covering data such as total population, average household size, in-migration in the previous five years, persons in employment, female activity rates, etc. The Ministry is currently preparing estimates of the costs involved in transferring the kilometre grid data held by the Study into a format suitable for LINMAP input.

The SYMAP project is still in its early stages. A number of trial maps of total population using kilometre grid data has been produced in order to decide the most appropriate format. Again a scale of 1 : 250,000 has been used. As a result of these preliminary tests several refinements have been suggested and are currently under investigation.

City of Coventry

In conjunction with the Ministry of Housing and Local Government, the Planning Office at Coventry have developed the Point Data System. The system holds information on every individual property and parcel of land in the city, approximately 120,000 records.

The basis of the system is the Gazetteer, consisting of the rating property number, the twelve-digit co-ordinate reference and the postal address. The existing computerised rating system provided a virtually comprehensive source of addresses whilst digitised co-ordinate references were obtained using a d-mac digitiser. To ensure compatability in merging information from these two sources it was necessary to pre-code the 380 Ordnance Survey 1 : 1250 sheets covering the city in a specific manner.

The system has been constrained in its development to one magnetic tape file – the present Corporation Computer is a LEO 3 – therefore the Gazetteer has been extended to include a further number of data fields, for example, Development Control monitoring information. The expanded Gazetteer, the Master File, is the overall control of the system.

Information relating to particular urban topics is held on manually operated topic files. Topic files have been developed for shop, office and industrial establishments.

Complementary to the system are 1 : 1250 up-dated Ordnance Survey base sheets. These sheets, having been up-dated to a base time – June 1969 – from an aerial survey,,are up-dated monthly in conjunction with the computer records. Monthly up-dating is achieved by tapping the existing Corporation monthly information flows. The topic files will be up-dated annually.

Appendices

Appendix 1

Catalogue of UK statistical data available for mapping

Introduction

The absence of any coherent catalogue listing the great wealth and variety of statistical data that exists about Britain seems surprising: it probably reflects the somewhat independent way in which the various agencies of central government have in the past collected the data of interest to them. This document is a modest attempt to bring together bibliographic details of certain of the data which seem likely to be of value for planning, and, more particularly, for mapping.

It is perhaps worth remarking at the outset that much of the data has been collected with the clear intention of presenting it in statistical rather than in cartographic form. The problems of geographical referencing consequently seem to have attracted less attention in the past than they do today when relationships between measured quantities and their location are the basis of computer mapping. Geographical referencing and the possible methods of carrying this out is a subject that has attracted our interest, and our conclusions on this subject are set out elsewhere in this report.

This catalogue deals for the most part with statistical data and it makes the reasonable assumption that the details of Ordnance Survey maps are known. It is self-evident that these maps form the base in relation to which economic and sociological patterns shown up by the statistics will be studied.

Any catalogue of this kind is inevitably out-of-date well before it is published. This one also suffers from the open-endedness of the subject: he would be a brave man who specified what data were and what were not of use for regional and local planning problems. However a list of essential data items was prepared in this connection by the East Anglia office of the former Department of Economic Affairs. This was supplemented by lists of data in regular use provided by the planning departments of Norwich City Council and Norfolk County Council. In addition, an examination was made of a number of planning studies, recreation studies and shopping studies. Government publications examined included the 1962 and 1968 Town and Country Planning Acts, the 1968 Coutryside Act, the report on the future of development plans prepared by the Planning Advisory Group, and various supplementary regulations and orders.

The resulting list of data items consists of about 15 "high priority" items, together with many other items that were considered to be of interest, but of less importance from a planning point of view. Certain items have been excluded either because they seemed not of major importance to planning (e.g.

some of the specialised Board of Trade enquiries), or because they are not geographically subdivided, or because the information is not available in a standard format for a large part of the country (e.g. specialised information collected by local authorities for local purposes).

Under the heading, "Geographical Resolution", a distinction is made between published material, data processed but not published, and basic data relating to individual information units. Processed data are usually in a reasonably convenient summary or machine-readable form, whereas basic data, that is, raw or near-raw data relating to the individual unit of collection such as the household, person, establishment or land parcel, are often in a relatively inconvenient form involving large numbers of questionnaires or files in a format not suited to machine sorting. Even if information is available for small areal units, and is not subject to confidentiality restrictions, this is no guarantee that there will be a detailed location code attached to the data item.

Basic information on machines and languages in use for processing has been obtained. For extra detail of tape format and program procedures that would be necessary for any further use of the data, it would be essential to undertake a depth study of selected items only, visiting data processing rather than survey staff.

Three tables precede the catalogue entries; Table 1 summarises some of the more important entries in the catalogue, and Table 2 summarises machine readability for some items. Table 3 indicates the comparability, stability and location source for boundaries used for data presentation.

Arrangement of entries in the Catalogue

1. Subject

Title of Material:	Official, or name in common use
Content:	Describes range and nature of subject matter
Geographical Coverage:	Included wherever possible, the area of coverage for the data item in question, not necessarily the whole area of responsibility of the originating body.

2. Method of data collection

Describes organisation of data collection, sampling procedures, and main stages of data processing.

3. Geographical resolution

Indicates the smallest geographical area for which information is available.

4. Statistical and legal restrictions

Statistical limitations:	If the survey has been conducted on a sample basis, this indicates the nature of the sample and the restrictions on its usefulness for mapping purposes.
Confidentiality:	Describes the limitations on the use of the data for mapping purposes in cases where there are safeguards against disclosure of confidential information (usually laid down by statute).

5. Boundaries

Gives the sources of boundaries for geographical areas for which the data were collected or (more usually) to which they have been aggregated.

6. Forms of data storage

Original data:	The number, type and organisation of files of raw data
Processed data:	The number, basic format and organisation of files of processed data are described.

7. Operational details

Originating body:

Location of supervisory staff:

Frequency and dates of survey:

Location of original data.

Location of processed data:

Publications:

Items covered

Table 1

Major Characteristics of Selected Items in Catalogue

Originating Department	**Ministry of Agriculture, Fisheries and Food**	**Ministry of Agriculture Fisheries and Food**	**Ministry of Housing and Local Government**
Title of material	Agricultural Census	Land Classification	Housing Statistics
Areal cover	England and Wales	England and Wales	Great Britain
Survey type/frequency	Annual (June)	Survey in progress	Monthly returns
Date of most recent census	1970		
Numbers involved	250,000 holdings	18 sheets published so far (on OS 1″ base)	95% housing progress is covered by returns received by 11th working day of the following month
Basic collection unit	Agricultural holdings	Variety of source information: NAAS, Soil Survey, Met. Office, field work, IGS	Local authority progress reports
Confidentiality restraints	3 holdings is minimum that can be disclosed		Most information is published
Statistical restraints	some minor exclusions in coverage		Delayed returns produce marginal time lag
Existing data aggregated nationally	Magnetic tape		
Existing data aggregated regionally	Magnetic tape	Much draft material available at regional offices	
Local data	Magnetic tape (Parish data if >2 holdings)		MSS (Local Authority progress report)

Originating Department	**General Register Office**	**Board of Trade**	**Board of Trade**
Title of material	Population Census	Census of Production	Census of Distribution
Areal cover	England and Wales	United Kingdom	GB (doesn't cover Isle of Man or Channel Islands)
Survey type/frequency	Basically decennial	Full census every 3 - 5 years. Sample in other years.	Full census roughly decennial + periodic sample censuses
Date of most recent census	Census 1961; Sample Census 1966	1968	1961 (full) + 1966 (sample)
Numbers involved	16 million schedules in Census 1961	130,000 establishments	1961: 300,000 returns 1966: 85,000 shop-keepers in sample
Basic collection unit	Households	Establishments	Retailers
Confidentiality restraints	Individual returns highly confidential	Data confidential if for group of 3 or fewer businesses	Individual establish-ment data highly confidential
Statistical restraints	1966 Census based upon 10% sample		1961: 88% response 1966: sample
Existing data aggregated nationally	Magnetic tape	Magnetic tape	Magnetic tape
Existing data aggregated regionally	Magnetic tape	Magnetic tape	Magnetic tape ("Old Standard Regions" in 1961)
Local data	Magnetic tape (Ward, parish and ED summaries)	Magnetic tape (Employment exchange)	Magnetic tape (Local authority for 1961; in 1966 for LA's if data complete)

Originating Department	**Ministry of Transport**	**Ministry of Transport**	**Ministry of Transport**	**Ministry of Transport**
Title of material	National Travel Survey	Survey of Road Goods Transport	Road Accident Statistics	General Traffic Census
Areal cover	Great Britain	Great Britain	Great Britain	Great Britain
Survey type/frequency	Periodically: ('64, '65, '66; next probable in 1971)	Periodically	As they occur	Periodically
Date of most recent census		1967/8		1965
Numbers involved	'64: 5,000 hsehlds. '65: 12,500 hsehlds. '66: 5,250 hsehlds. '71: ? 15,000 "	1967/8: 49,000 vehicles		2,200 census points
Basic collection unit	Households	Vehicles	Accidents	Vehicles at census points
Confidentiality restraints		Individual returns confidential		
Statistical restraints	Sample	Sample		Sample
Existing data aggregated nationally	Cards 1964 and 1966	Magnetic tape	Magnetic tape	Magnetic tape
Existing data aggregated regionally	Cards - 1965	Magnetic tape (MoT Standard Zones)	Magnetic tape	Magnetic tape
Local data			Magnetic tape (individual accident information available)	Magnetic tape(main trunk link data) MSS (census point tally books)

Originating Department	**Department of Employment & Productivity**	**Department of Employment & Productivity**	**Department of Employment & Productivity**
Title of material	Register of Un-employed Persons	Register of Employers (ER 1)	Local Employment Records (ER 2)
Areal cover	Great Britain	Great Britain	Great Britain
Survey type/frequency	Monthly estimates	Updated from official returns and local knowledge	Annual
Date of most recent census			1970
Numbers involved		About 1,000 em-ployment exchanges	About 1,000 employ-ment exchanges
Basic collection unit	People	Employers within employment exchanges	Employees within employment exchanges
Confidentiality restraints	Individual data highly confidential	Available only in special circumstances to central and local government	Individual establishment data confidential, except to government
Statistical restraints	Only covers reg-istered unemployed who seek work	Not intended to be a comprehensive register	Excludes some cate-gories of employees
Existing data aggregated nationally			
Existing data aggregated regionally			
Local data	MSS (Employment Exchange)	MSS (Employment Exchange)	MSS (Employment Exchange)

Table 2

Forms of Automatic Data Storage for Selected Items

	Basic data units (unprocessed)	**Basic data units (machine processed)**	**Data aggregations**
Agricultural Census	Postal returns for individual holdings (available for 1960,1966 1967, 1968; about 250,000 cases).	Standard punched cards for each holding (for 1960, 1965,1966,1967,1968; about 750,000 cards for EE Deuce machine. From 1967 onwards, on magnetic tape for ICL 1907).	Printouts for parishes (about 12,000), National Agricultural Advisory Service Districts, and counties.
Census of Production	Postal returns for individual business establishments (about 130,000 cases in 1963).	Returns for each establishment on magnetic tape (for LEO 3).	Printouts by kind of business and exchange area.
Census of Distribution	Postal returns for individual retail establishments or groups (about 300,000 returns for 500,000 establishments for 1961).	Returns for each establishment on magnetic tape (for LEO 3).	Printouts by local authority area.
Survey of Road Goods Transport	Postal returns for sample of individual vehicles (about 40,000 for 1962, 45,000 for 1967/8).	Hollerith cards for individual vehicles (about 40,000) and trips (about 700,000 for 1962). 1962 material sorted on Hollerith machine. 1967/8 material on cards and magnetic tape.	Total accumulated for vehicle and trip characteristics. For 1967/8 material edited magnetic tapes show interzonal movements. (Analysis in progress)
National Travel Survey	Travel record diaries and questionnaires for households have been destroyed.	Information for individual households on cards and magnetic tapes. (CDC machine for 1964/5, UNIVAC 1108 for 1966).	Parameter tapes and analysis printout.
Population Census 1961 & 1966	Individual household schedules (about 16 million for 1961, 10% sample in 1966).	Individual household data on magnetic tapes (IBM 705 for 1961 & 1966, ICL 1904 for 1971).	Ward library summary on magnetic tapes (include ED data).
Local Employment Estimates (ER 2)	Returns for each establishment and for employees working elsewhere held (for a while) at local offices. Exchange summary forms and "debit/credit" forms held for 5 years (1,000 exchange forms, 41,000 "debit/credit" forms).	Processed data for individual establishments do not exist. SIC data for each local office held on magnetic tapes. Also "debit/credit" cards and tapes (EMIDEC 1100 for 1965-67; ICL 1900 series for 1969 onwards).	Printouts by local employment office.

Table 3

Boundaries – Comparability, Stability and Location Source

Type	Stability and Comparability	Source of Location Details	Surveys in which used
Enumeration Districts	Usually subject to many alterations between censuses. ED's for 1966 generally larger than those for 1961, but ED's for 1971 likely to be smaller than those for 1961. Aggregated to form wards and parishes.	General Register Office, but some local authorities have key maps from GRO source.	Population Census
Major Shopping Centres	Defined by Board of Trade for 1961 Census of Distribution. Extended as necessary for further censuses, but with more detailed system of location coding. Any similarity with other boundaries purely coincidental	Mapped by MHLG (mainly at 1 : 2,500 scale) and held at MHLG map library.	Census of Distribution
Ward and Civil Parishes	Subject to periodic and selective revision, but relatively stable. Only a small proportion of East Anglia parishes have been subject to revision since 1951. Wards and parishes are aggregated to form local authorities.	County administration diagrams at ½" scale (and more recently at 1 : 100,000 scale) show local authority and civil parish boundaries. MHLG hold master sets, including details of all changes. The library also holds a set of 2½" maps showing ward boundaries and changes. There is also a key showing which 2½" sheets have been subject to changes, and a set of 1" maps showing wards and civil parishes for 1961.	Population Census Agricultural Census (Civil parishes only).

Table 3 (continued)

Type	Stability and Comparability	Source of Location Details	Surveys in which used
Local Authorities	These boundaries are subject to occasional local changes or full-scale reorganisation such as occurred in London in 1965. Local authorities are aggregated to form counties (excluding County Boroughs) and conurbations as defined by the GRO.	Local authority boundaries are shown on County Administration Diagrams published periodically as changes occur. Also shown on one of the 10-mile series maps.	Population Census Census of Distribution (counties and towns exceeding 50,000 only) Housing Statistics Epitome of Accounts Rateable Value
Employment Exchange Areas	These are fairly stable boundaries, subject to occasional small changes and amalgamations. Employment exchange areas include local and branch offices, and may be amalgamated into "travel to work" areas for unemployment calculations. Boundaries (at least in East Anglia) do not normally coincide with local authority or county boundaries. Combinations of exchanges form employment districts.	Employment exchange boundaries have been plotted onto a series of County Administration diagrams, and are held at the MHLG map library. Changes in boundaries are noted and recorded, and summary maps at 10-mile scale are available for selected years.	Department of Employment and Productivity Enquiries
Administrative Counties and County Boroughs	See "local authorities". Combinations of counties and county boroughs from Ministry of Transport standard zones, and Economic Planning Regions.	See "local authorities".	Inland Revenue Income estimates (not county boroughs) Population Census (summaries) Agricultural Census (summaries) (not county boroughs) Vehicle Registrations Road Accidents

Table 3 (continued)

Type	**Stability and Comparability**	**Source of Location Details**	**Surveys in which used**
Ministry of Transport Standard Zones	Defined for Ministry purposes in the mid-1960's. Normally combinations of local authorities.	Ministry of Transport zone boundaries are obtainable from the Ministry.	Ministry of Transport Surveys, including Survey of Road Goods Vehicles.
Economic * Planning Regions	Combinations of county and county borough areas. Subject to periodical review. EPR's succeeded "Old Standare Regions" in use until mid-1960's.	Boundaries defined for use in various survey reports, e.g. Population Census, unemployment rates, incomes, etc. Available from many published sources.	National Food Survey National Travel Survey (except East Anglia) Census of Production. Many other statistics are amalgamated up to regional level, and in many cases accuracy checks and amendments are made at a regional level.

* Other "regions" in use include GPO regions, electricity, gas and British Rail regions, etc.

1. Subject **Agricultural Census**

Content: The Agricultural Census collects information on crop acreages, numbers of livestock and numbers of workers on farm holdings. The "occupier and his wife" and school-children are not included in employment figures, but farmers/directors/partners will be included from June 1970. Areas are to the nearest ¼ acre. The census provides a "snap-shot" situation for a particular date, e.g. 4th June 1969. Categories of animals, poultry and crops are given in considerable detail for June (about 170 categories) but simpler groupings are used for the sample census (about 75 categories).

Geographical coverage: England and Wales only

2. Method of Data Collection

The agricultural census is carried out under the Agriculture Act, 1947. Questionnaires are sent by post. The record of holdings is amended through notifications by local officials of MAFF, and by information from farm proprietors. The full census takes place in June each year. For the three sample censuses held in March, September and December, a random sample of one-third of all parishes is taken, and within these parishes, every holding within the scope of the Census is enumerated, so that each parish is sampled only once in addition to the full coverage of the June Census. The procedure has now been standardised so that the same group of parishes receives sample census questionnaires at the same time each year. The sample could be reduced in size from the present 80,000 holdings per quarter to about 30,000 holdings without loss of reliability if stratification were introduced. Experiments are currently in hand with a view to introducing stratified samples from March 1971. Sample data are simply aggregated to county level, and these are grossed up to represent the parent population at national level. There is, therefore, no tabulation of sample census data below a national level. In the grossing-up procedure, the latest June figures based on a full census are used to calibrate the grossing-up factors.

In the case of a holding which straddles a parish boundary, it is allocated to the parish in which the greater part of its acreage lies, regardless of the situation of the farm buildings. (This will result in under- and over-statement for the parishes concerned.) This dates back at least to the late 1950's, before which the decision was left to the discretion of the divisional office, and there was apparently a tradition that the situation of the farm buildings was the main criterion.

Where there are multiple holdings (i.e. holdings owned or occupied by the same farmer), separate returns are rendered for each holding and each bears the code letter "M". If these holdings are amalgamated for statistical purposes, (which is the practice where holdings are run as a single farm unit), so that one return is rendered for them all, then it would normally be to the home farm that the census form would be sent. But this could be varied by arrangements between the divisional office and the farmer.

Notifications of parish boundary changes are received from the Ministry of Housing and Local Government as and when they occur, and they are passed on to the divisional offices concerned to obtain particulars of the holdings affected; later, after any amendments to the parish list have been completed, the regional office is notified. At the June census date, the Ministry's parish boundary records are as up-to-date as possible, but changes notified too late for all the amendment procedures to be carried out in time would not, of course, be included; but they are taken into account the following year. These changes are not regarded as a serious drawback as time-series analysis is not usually undertaken at a parish level.

Returns are required from all agricultural holdings over one acre in size, excluding 50,000 holdings regarded as "statistically insignificant". The criteria of significance are (a) the level of labour requirement in standard man-days and (b) acreage. (Actual criteria: less than 26 standard man-days, under 10 acres of crops and grass and no regular full time workers.) Prior to 1968, less rigorous criteria were used, and only 13,000 holdings over one acre were excluded, for which allowance was made in the published results. (Allowance is no longer made for these statistically insignificant holdings which are now entirely excluded from the census.) As a corollary to the exclusion of these SI holdings over one acre, however, it is proposed to bring into the census as from June 1970 holdings which do not have more than an acre of land used for agriculture, but which nevertheless have a significant output of agricultural produce. The types of holding which will thus be brought in are particularly those with intensive forms of production in horticulture and pig and poultry farming.

3. Geographical Resolution

(a) Basic Information Unit: Pre-1960: All census returns destroyed

Annual June census returns for individual farms (subject to confidentiality restrictions — see 4. below). Coded by county/parish/ holding number.

Sample censuses — returns for individual farms (subject to confidentiality restrictions — see 4. below). Geographical resolution restricted by level of sample reliability.

(b) Processed data: Pre-1960: All censuses — parish totals only

Annual June census — individual farm data and parish summaries (data for individual farm subject to confidentiality restrictions — see 4. below).

Sample censuses — individual farm data (subject to confidentiality restrictions — see 4. below). Geographical resolution restricted by level of sample reliability. Parish summaries.

(c) Published material: Pre-1960 (all censuses) and annual June census — county summaries.

Sample censuses — grossed up to national level.

4. Statistical and Legal Restrictions

Statistical limitations: Sample censuses are only used at the national level and are grossed up by reference to the most recent June census. The results of the latter are subject to three types of error:-

a coverage of holdings of insignificant size or output (see 2. above);

b response — generally excellent; and

c locational — results for small areas may be over- or under-estimated as a result of the methods for allocating holdings to parishes (see 2. above).

Confidentiality: The terms of the Agricultural Act, 1947 (Section 80) prohibit the disclosure of any information from individual returns except with the consent of the informant, or in the special circumstances specified in that Section of the Act. There is an unwritten rule that none of the statutory provisos is invoked if there is a risk that the information disclosed may be used to the farmer's disadvantage.

Parish summaries of census data are thus freely available, subject only to the following restriction. Where there are only one or two holdings in a parish, irrespective of their size, a summary is not provided. Instead the data are amalgamated with that of a single adjoining parish. Where there are three or more holdings in a parish, no suppression occurs.

5. Boundaries

Local authority and parish boundaries are available on county maps published by HMSO. These maps are up-dated periodically. More recent changes are recorded at Ministry of Housing and Local Government map libraries.

6. Forms of Data Storage

Original data: Returns are obtained from about 250,000 holdings over one acre in size, excluding 50,000 holdings regarded as "statistically insignificant". Each return has an address and a county/parish/holding code. The acreage of the holding is given, but the location of the land is not specified. All that can be deduced is that the majority of the land referred to is located somewhere in the parish in question. Returns more than about five years old are destroyed, except for 1960 which was a "World Census of Agriculture" year. In future, data for the years ending in -0 will be preserved for 25 years.

Processed data: All data from returns are punched on standard 80-column cards, about 13 answers per card, 3 cards per return, making 750,000 cards altogether. This is likely to reduce to perhaps 500,000 in future years as the number of holdings declines, and as a more economical card layout is introduced. From 1960 to 1967, cards were sorted and then analysed on an English Electric Deuce computer, reading machine code only. For the 1967 census records, and ICL 1907 computer has been used, creating magnetic tapes for holding data, and using the COBOL programming language. Tapes are arranged by holdings within parishes within counties. New programs and card layouts are being used for the new computer, so there is no opportunity at present for machine comparisons between pre- and post-1967 data.

Both punched cards and tape include the county/parish/holding code for each holding. Summary coded print-outs are prepared for major topics, holding by holding. Items are accumulated for counties, and these are aggregated (and grossed-up in sample censuses) for MAFF regions and England and Wales as a whole. (MAFF Eastern Region is the East Anglia Economic Planning Region, plus the whole of Essex, Hertfordshire, Bedfordshire, parts of Greater London and parts of Holland (Lincolnshire). Figures for the Economic Planning Region can be obtained by aggregating the county summaries, so this lack of fit is only an incidental problem.) Summary sheets are also prepared for parishes and NAAS districts, which are groups of several parishes. There are about 12,000 parishes in England and Wales, but in about 500 cases it has been necessary to amalgamate two adjoining parishes because of confidentiality problems. Key maps showing the boundaries of parishes are held at Guildford, and there is also an alphabetical register of parishes giving the grid reference of the 10-kilometre square in which the parish occurs.

7. Operational Details

Originating body: Ministry of Agriculture, Fisheries and Food

Location of supervisory staff: Agricultural Census Branch,
Government Buildings,
Epsom Road,
Guildford, Surrey.

Frequency and dates of surveys/censuses: Full agricultural census takes place every June. In addition, there are sample censuses, each covering one-third of all holdings, in March, September and December. Thus, each holding coming within the scope of the census provides two returns per year — see 2. above.

Location of original data: All pre-1960 data for individual holdings have been destroyed, and data for that period are available only in parish summary form. June census returns for 1960 and every following tenth year are retained indefinitely; those for even years (other than 1960, 1970 etc.) are retained for five years; those for odd years, for three years. The June forms for 1960, 1966, 1967, 1968, 1969 are therefore still available and stored in the MAFF file store at Hayes; those for the intermediate years have been destroyed. (See attached schedule of location/destruction.)

Location of processed data: All pre-1960 data for individual holdings have been destroyed, and data for that period are available only in parish summary form. Punched cards for 1960 and every tenth year, and those for 1965, are retained indefinitely; for every tenth year after 1965, for ten years; and for intermediate years, for five years. The June Census data cards for 1960, 1965, 1966, 1967, 1968 and 1969 are therefore still available – the first three at Hayes, the last three at Guildford. 1967, 1968 and 1969 data held on magnetic tape will be kept for five years, at Guildford. (See schedule.)

Publications: Apart from the parish and NAAS district summary records which are held in book form at Guildford, county, regional and national figures are published annually in "Agricultural Statistics", including cross-analyses by type of farm, size of holding, labour, etc. Advance statistics on a one-third sample basis are drawn from the June census pending the publication of the full results.

Location/Destruction Schedule

MAFF June Census	Original Returns	Cards	Magnetic Tape
Pre 1960	destroyed (parish summaries only)		
1960	Hayes (will be retained indefinitely)	Hayes (will be retained indefinitely)	
1961	destroyed (were kept for three years)	destroyed (were kept for five years)	
1962	destroyed (were kept for five years)	destroyed (were kept for five years)	
1963	destroyed (were kept for three years	destroyed (were kept for five years)	
1964	destroyed (were kept for five years)	destroyed (were kept for five years)	
1965	destroyed (were kept for three years)	Hayes (will be retained indefinitely)	
1966	Hayes (will be kept for five years)	Hayes (will be kept for five years)	
1967	Hayes (will be kept for three years)	Guildford (will be kept for five years)	Guildford (will be kept for five years)
1968	Hayes (will be kept for five years)	Guildford (will be kept for five years)	Guildford (will be kept for five years)
1969	Hayes (will be kept for three years)	Guildford (will be kept for five years)	Guildford (will be kept for five years)

1. Subject **Agricultural Land Classification**

Content:

The Land Classification Survey was designed to assist the Agricultural Land Service in advising on the release of agricultural land for urban development. Maps in the series are provisional and are prepared on a scale of 1 : 63,360. The agricultural land is graded, on a national basis, according to the degree to which its physical characteristics impose long-term limitations on its agricultural use. The main physical factors which have been taken into account are climate (particularly rainfall, transpiration, temperature and exposure), relief (particularly slope) and soil (particularly wetness, depth, texture, structure, stoniness and available water capacity). Ability to grow a wide range of crops (including grass), whether actual or potential, is given considerable weight but it does not outweigh the ability to produce consistently high yields of a somewhat narrower range of crops. The major non-agricultural areas are also shown. They are divided into two classes – (a) existing urban areas and land scheduled for urban use and (b) other areas primarily in non-agricultural use, e.g. woodland and large public open spaces outside the built-up areas.

Geographical coverage: England and Wales only.

2. Method of Data Collection

The classification draws upon material provided by the Soil Survey, National Agricultural Advisory Service, Meteorological Office, Institute of Geological Sciences, and fieldwork by staff of the Agricultural Land Service Research Group.

3. Geographical Resolution

(a) Basic Information Unit: Information assembled from a variety of sources: Soil Survey, NAAS, Meteorological Office, Institute of Geological Sciences, field work. Varies in level of resolution.

(b) Processed data: Five categories of agricultural land and two categories of non-agricultural land mapped on a scale of 1 : 63,360; normally no area of agricultural land of less than 200 acres in extent is separately graded.

(c) Published material: Maps published at 1" scale: five categories of farmland identified.

4. Statistical and Legal Restrictions

Statistical limitations: The only statistical limitations which apply relate to the scale of the maps. A survey on a scale of 1 : 63,360 is essentially of a reconnaissance nature, and any enlargement of the maps could be dangerously misleading. (See also 6. below.)

Confidentiality: No restriction on published data; pre-publication material could be made available where sufficiently advanced.

5. Boundaries

The five land classification categories are marked in colour on to OS 1" to the mile outline maps — available as published by MAFF.

6. Forms of Data Storage

Original data: Using the sources mentioned in 2. above, Land Classification data are assembled in map form at 1" to the mile scale. These are held at the regional offices where the research officers are based.

Processed data: Land classification information is published as a series of colour categories superimposed on normal 1" OS outline maps. The five grades mentioned above are shown, together with major urban areas (in red) and other areas outside built-up areas in non-agricultural use,e.g. woodland, public open space (in orange). Maps are only published when land classification information for an entire OS sheet becomes available, and publication is therefore lagging behind the progress of field work. About half the country has been surveyed, but only eighteen sheets have been published. Work is tending to concentrate on areas where there is pressure for development of farmland. It is stressed that the land classification map is a macro-scale reconnaissance document, and that lines on the map will normally represent zones of transition on the ground. Where very mixed conditions occur, an assessment is made of which grade predominates or is most typical; this grade is then assigned to the whole area. Unless divisions between grades are determined by obvious physical features, no area of less than about 200 acres is separately graded.

7. Operational Details

Originating body: Ministry of Agriculture, Fisheries and Food

Location of supervisory staff: Agricultural Land Service,
Great Westminster House,
Horseferry Road, London, SW1.

Frequency and dates of survey: The initial aim is to prepare provisional land classification maps for the whole of England and Wales at 1" scale. Work has been in progress for about two years, and approximately half of the country has been covered, although only eighteen sheets have been published.

Location of original data: At MAFF regional offices.

Location of processed data: Published in map form at a scale of 1 : 63,360 on OS base maps. Available from Ministry of Agriculture, Fisheries and Food, (Publications), Government Buildings, Tolcarne Drive, Pinner, Middlesex HA5 1 DT.

Publications: See 6. above. There is an explanatory note covering the published maps, and Technical Report No. 11 of the Agricultural Land Service (MAFF, 1966) describes the purpose and methodology of the Land Classification in detail.

1. Subject **Farm Management Survey**

Content: The Survey provides data about the cropping and stocking of the sample farms and financial data relating to their output, inputs and net income. These can be related to region and the size and type of farms.

Geographical coverage: England and Wales

2. Method of Data Collection

The survey is carried out by University Departments of Agricultural Economics which collect and analyse a sample of farm accounts in each province. The sample used is small (about 2,500) and not randomly selected, since it is voluntary, unpaid and chosen for willingness to provide accounts. About 10% of the farms in the sample change on average every year. The effect of disproportionate representation for most of the types and sizes is corrected by weighting the averages by weights from the Agricultural Census, but sufficient accounts for adequate representation of the horticultural and poultry industries are not obtainable. Returns sent by the Universities to MAFF headquarters are scrutinised before being sent to Guildford for processing, during the course of which credibility checks are carried out by the computer.

3. Geographical Resolution

(a) Basic information unit: Name, address, 10-kilometre square reference of each farm in the sample. Geographical resolution restricted by level of sample reliability.

(b) Processed data: County number and 10-kilometre square reference. Geographical resolution restricted by level of sample reliability.

(c) Published data:
(i) University reports present results analysed in terms of "farming districts" within their provinces – without defining these districts or discussing the level of sample reliability.
(ii) MAFF report amalgamates data for counties into five large groupings designed to distinguish the broad regional differences.

4. Statistical and Legal Restrictions

Statistical limitations: For reasons of cost, the sample size is small and not randomly selected (see 2. above for more detail). This limits the amount of statistically reliable results, though year to year changes in financial position are thought to be fairly indicated. Average data of net income, relating to farms of a particular type and size, are also regarded as reasonably representative.

Confidentiality: The individual returns are confidential, and the information collected is made available by farmers on the clear understanding that their names and addresses will not be divulged. For this reason specific location codes are not part of the return as sent to MAFF. Any controlled access to processed data would require a minimum of ten farm accounts within a type of farming cell, by size of business by region.

5. Boundaries

"Farming districts" as used by the Universities are not precisely defined. Regional boundaries adopted by MAFF are presented in their report.

6. Forms of Data Storage

Original data: The various university departments collect the data from the individual farms in the sample, and at this stage it is clearly possible to specify farm location. When the returns are sent to MAFF headquarters, only two locational references remain: (a) the county number and (b) a 10-kilometre grid square reference.

Processed data: Data from individual returns are transferred to punched cards at Guildford. The cards carry the county number as a locational code. Since 1966 the data have been transferred to magnetic tape, and for the 1969 survey there will be some use of discs.

7. Operational Details

Originating body: University Departments of Agricultural Economics and the Ministry of Agriculture, Fisheries and Food.

Location of supervisory staff: The University Departments in each province (see list in MAFF publication mentioned in this section) and Economics Division 1, MAFF, Whitehall Place, London SW1.

Frequency and dates of survey: Each year since 1936-37. The earliest years were covered by reports issued by the Agricultural Economics Research Institute at Oxford.

Location of original data: Returns are retained by the individual university departments carrying out the survey; MAFF retains copies of these at HQ (and Hayes). Duplicate copies are sent to the Agricultural Economics Research Institute at Oxford.

Location of processed data: MAFF, Guildford

Publications: Each University Department responsible for part of the survey usually publishes its own report. MAFF publishes a report, "Farm Incomes in England and Wales", which makes analyses in terms of five large groupings of counties, and compares the results with those of the identical sample of the previous year.

1. Subject **Farm Improvement Schemes under the Agriculture Acts 1957 and 1967**

Content: There is a considerable range of improvements for which farmers (owners or tenants) can obtain grants. However, claims are made on a standard form, the authorising office having the duty of classifying the proposed improvement.

Geographical coverage: England and Wales.

2. Method of Data Collection

Claim forms are sent first to divisional offices for approval, and are then forwarded to Guildford for central payment and statistical analysis.

3. Geographical Resolution

(a) Basic information unit: Individual farm holding.

(b) Processed data: Individual farm holding.

(c) Published material: Not applicable.

4. Statistical and Legal Restrictions

Statistical limitations: Whilst information is collected about all investments aided by the Farm Improvement Schemes, it is fair to point out here that only a proportion of total farm investment receives such grant aid. For all farm investment (excluding houses) this proportion is about 50%, but it varies according to the size and type of farming, and the type of farm investment.

Confidentiality: Individual applications for grants are treated as confidential. If data were required for research purposes for areas smaller than counties, then some data amalgamation would probably be necessary to avoid disclosure about individual applications.

5. Boundaries

Not applicable.

6. Forms of Data Storage

Original data: The application forms describe the applicant (tenant, owner occupier etc.) and the type of improvement proposed. Type of farming is obtained from information derived from census returns.

Processed data: The claim forms are processed at Guildford, the first item being the county/parish/holding number. Also coded is the type of improvement, type of applicant, and from 1st April 1968 onwards, the type of farming (eight categories).

7. Operational Details

Originating body: Ministry of Agriculture, Fisheries and Food.

Location of supervisory staff: Ministry of Agriculture, Fisheries and Food,
Economics Division,
Whitehall Place,
London, SW1.

Frequency and dates of survey: This is not a survey, the data being a by-product of an administrative process. The data are therefore a continuous flow, but could be tied to the date of payment. The original scheme began operation in 1957.

Location of original data: Application and claim forms for grants are sent to divisional offices. A number of administrative levels is involved in the authorisation process, but the original forms are filed at the divisional level.

Location of processed data: Claim forms are sent to Guildford when grants have been authorised; there they are processed for payment and then returned to Divisional Offices.

Publications: None. (However, cumulative progress reports are produced each month. Information from these can be given to members of the public on request, but they are mainly intended for Ministry staff.)

1. Subject **(Agricultural) Wages and Employment Enquiry**

Content: The survey provides data about agricultural workers in terms of their main occupations, hours of work, earnings in cash and in kind; a number of cross tabulations can be made which relate these data to Economic Planning Regions, occupation and labour size group.

Geographical coverage: England and Wales

2. Method of Data Collection

Enquiry forms are filled in by Ministry inspectors as they visit agricultural and horticultural farm holdings within the sample (see below). Wages Inspectors have a statutory duty to ensure that agricultural workers are receiving the legal basic minimum rates of pay. The Wages and Employment Enquiry itself is designed as a "spot check" to ensure that workers are not being underpaid, and to collect reliable estimates of average earnings and hours. One report form (AW 4) refers to the farm, whilst other forms

(AW 3) refer to the individual farm workers.

Sampling method: The sample for each year is selected from the most recently available June Agricultural Census. A magnetic tape is prepared which includes every holding in England and Wales, arranged in county/parish/holding number order, and giving all the census details. A stratified random sample is picked for the Enquiry annually. The stratification is by nine labour size groups. Because of the way the census information is ordered on the tape (viz. firstly by counties, then by parishes and finally by numerical sequence of holding numbers within parishes), the sample when it is drawn is stratified by labour size group and is proportional to the number of holdings in any one county. Number of holdings in the sample is about 4,500, or between 4½% and 5% of those holdings having labour. From June 1970 onwards, the sample will only be drawn from holdings having hired labour, rather than from all holdings having labour (including family workers) as at present.

3. Geographical Resolution

(a) Basic information unit: County/code number; address of the farm; address of the farm worker. Geographical resolution restricted by level of sample reliability.

(b) Processed data: County/code number of the farm holdings where the workers are employed. Geographical resolution restricted by level of sample reliability. (Provisional figures for Economic Planning Regions have been produced, starting 1967.)

(c) Published material: Refers to Great Britain as a whole (incorporating data from the Department of Agriculture and Fisheries for Scotland), or to England and Wales only.

4. Statistical and Legal Restrictions

Statistical limitations: For national tabulations the majority of the data have a sampling error of less than 1%. The most reliable regional information produced is that referring to the level of earnings and hours for all hired men. Data relating to different occupations are not reliable at this level.

Confidentiality: Individual returns are treated as confidential, especially since they contain information collected statutorily under the Agricultural Wages Act, 1948. Unpublished data (for England and Wales and provisional results for Economic Planning Regions) are available to interested bodies on application.

5. Boundaries

Economic Planning Regions are available from HMSO.

6. Forms of Data Storage

Original data: The report form AW 4 contains general information about the farm, and specific information about the numbers and types of workers; it includes the name of the occupier and the address of the farm. The report form AW 3 contains the basic information about earnings and hours, type of work, and includes the worker's name, the name of the employer, and sometimes the employee's address.

Processed data: All data transferred to punch cards monthly. From 1968 onwards, limited data on the magnetic tape quarterly. Cards are retained for about five years at Hayes, Middlesex. The processed data retain the county code number.

* From April 1970 onwards, all data to be transferred to magnetic tape.

7. Operational Details

Originating body: Ministry of Agriculture, Fisheries and Food.

Location of supervisory staff: Economic Division 1,
Ministry of Agriculture, Fisheries and Food, West Block,
Whitehall Place,
London, SW 1.

Frequency and dates of survey: The survey is carried out continuously each year, and began in October 1945. The basic time unit is each quarter's results, these results being built up from monthly observations, although individual months' results are not reliable enough for use. The calendar year used for the internal reporting of the survey ends on 30th September of each year.

Location of original data: Questionnaires are retained at Divisional Offices for varying and indeterminate periods (probably on average from three to five years).

Location of processed data: Punched cards retained at MAFF, Hayes, Middlesex for five years after they have been analysed. From April 1970 onwards, data will only be on magnetic tape and retained at MAFF, Guildford.

Publications: (a) A Press Notice setting out average weekly earnings and hours of agricultural workers in England and Wales is published quarterly.

(b) The Employment and Productivity Gazette:-

Monthly: An index of the average weekly earnings of agricultural workers in England and Wales.

Twice yearly: The average weekly earnings of men, youths and females in Great Britain for half-yearly and yearly periods. The average hours worked and hourly earnings of men, youths and females in England and Wales for half-yearly and yearly periods.

Annually: A comprehensive article setting out details of average weekly earnings, hours, payment in kind etc. by occupations of men, youths and females in Great Britain over the year.

(c) Annual Abstract of Statistics:

The average weekly earnings of men, youths and females in Great Britain for half-yearly and yearly periods.

(d) In addition the following are available on request:

Quarterly: (i) Detailed statement of hours and earnings by occupations of men, youths and females in England and Wales.

(ii) Hours and earnings of men by Economic Planning Regions.

Twice yearly: (iii) Statement as at (a) for years ending March and September.

Annually: (iv) Annual report of the Wages and Employment Enquiry.

(e) A summary of the main results of the Enquiry for the period 1945-65 was published in 1967, entitled "The Changing Structure of the Agricultural Labour Force in England and Wales" (MAFF, 17s.6d.).

1. Subject **National Food Survey**

Content: Survey covers samples of private households in Great Britain, and obtains details of household characteristics, food consumption, expenditure, prices and nutrition. Until January 1950, main survey confined to urban working class households – thereafter extended to all classes and all parts of Great Britain.

Geographical coverage: Great Britain

2. Method of Data Collection

The Survey covers about 8,000 households a year. In a specially supplied log-book, the housewife records the description, quantity and cost of all food which enters the household supply during a period of one week. The survey excludes vitamin preparations, food bought exclusively for pets, sweets, chocolates, soft drinks, alcoholic beverages. The housewife makes separate notes of meals consumed out of home by number and type but not by composition or cost. She also records quantity of milk supplied to her children under the School Milk Scheme. The interviewer visits the household three times, (or more if difficulties arise) and obtains additional information about the household characteristics, e.g. income of head of the household and of the family.

Sampling techniques: Three-stage stratified random sampling scheme. Sampling frame covers the whole of Great Britain. The stages are as follows:

(i) selection of parliamentary constituencies,
(ii) selection of polling districts within the chosen constituencies
(iii) selection of households within polling districts.

Note: For further details of methodology — see the Annual Reports of the Food Survey Committee.

3. Geographical Resolution

Basic information unit: Log-books for individual households (subject to confidentiality restrictions — see 4. below). Geographical resolution at a point in time restricted by level of sample reliability, but this limitation is less applicable to averages taken over a period of years.

Processed data: Data for individual households (subject to confidentiality restrictions — see 4. below). Geographical resolution at a point in time restricted by level of sample reliability, but this limitation is less applicable to averages taken over a period of years. Summaries for region and type of area, and some brief information for parliamentary constituencies are included in the sample.

Published material: Region and type of area (London conurbation; provincial conurbation; urban areas (larger towns); urban areas (smaller towns); semi-rural areas; rural areas.

4. Statistical and Legal Restrictions

Statistical limitations: The results obtained are subject to sampling error, but this is not normally more than two or three times the standard error. Estimates of standard errors are included from time to time in the annual reports.

Confidentiality: The original NFS log books are not available to anyone outside MAFF or the Government Social Survey because a specific assurance on the point has been given to participants. Access to computerised records is, in principle, possible subject to a sufficient degree of aggregation (at present under review) to ensure that the pledge of confidentiality is not infringed. Access to available summarised records is permitted to bona fide research workers and public bodies. Commercial firms can be provided with summarised results upon payment of a fee.

5. Boundaries

Standard maps for old standard regions (applicable to analyses of data prior to 1967) and New Economic Planning regions (applicable to data from 1967 onwards).

6. Forms of Data Storage

Original data: Household log-books recording description, quantity and cost of all food entering the household for seven consecutive days. Log-books do not show the name or postal address of the respondent.

Processed data: The basic data are currently punched on to paper tape for analysis by an Elliot 803 computer. Processed data retains location codes, which identify regions (formerly old standard regions, now Economic Planning regions), type of area (according to degree of urbanisation) and parliamentary constituency. The tabulations provide quarterly and annual averages of household food purchases, expenditure, prices, free supplies and energy value and nutrient content for Great Britain and for households of different composition and income grade, and for standard regions of England, for Scotland and for Wales and for various types of area classified according to degree of urbanisation.

7. Operational Details

Originating body: Ministry of Agriculture, Fisheries and Food

Location of supervisory staff: MAFF,
Tolcarne Drive,
Pinner, Middlesex.

Frequency and dates of survey: Survey carried out continuously week by week since 1940, except for short breaks. Reports made annually.

Location of original data: See Location of supervisory staff.

Location of processed data: Paper tape retained for a limited period by MAFF, DPD, Guildford.

Publications: Annual reports of the National Food Survey Committee – Household Food Consumption and Expenditure (HMSO). In 1967, this had 50 tables (with 24 more in the appendices and supplement) giving details of various aspects of food consumption and expenditure. Figures given for nine regions and six types of area.

Other Ministry of Agriculture, Fisheries and Food, Census Branch, Enquiries Undertaken by Staff at Guildford.

In addition to the June census and the quarterly samples that duplicate it, there are a number of additional enquiries carried out by the Census Branch staff. These cover specialist agricultural activities in fuller detail, taking the form of a follow-up questionnaire sent to holdings falling in certain categories on the basis of June census results. There is a "vegetable enquiry" in September and December involving all holdings where vegetables for human consumption were reported in the previous June. There is a similar glasshouse census in January and July. In every second year there are special enquiries into spray irrigation and glasshouse equipment, and an orchard enquiry every four years. Questions about agricultural machinery are included with the three sample censuses that supplement the June census, but are not included on the June census forms.

The main June census is used to provide sampling frames for a number of specialised sample enquiries into particular agricultural activities. Estimates of the pig population are made every six weeks by Ministry Field Officers who visit a random sample of about 2,000 farms. The sample is changed every three months. Information is also collected about egg yields, systems of management and egg disposal by means of a field survey of a changing monthly sample of 400 egg producers. (Recently this survey has been extended to cover prices of direct sale eggs, and at the same time certain changes were made in the structure of the sample, entailing collection of some of the information by post.)

A sample of about 2,900 farms is used each year to estimate the stocks of unsold cereal held at the end of each month during the grain marketing season. This enquiry is carried out partly by post and partly by visits. There is a further sample of 10,000 holdings at the end of the cereal year to obtain estimates of yields of the main cereal crops. These returns are compared with estimates of yields provided on the basis of about one-third of parishes by District Agricultural Advisers. Returns are sent to Guildford by County Agricultural Advisers where the results are checked and grossed up. Final estimates are on a county and national basis.

1. Subject — **Register of employers**

Content: Each employment exchange maintains, for administrative purposes, a register of the more important or principal employers in its area. The information recorded consists of name, address, industrial classification (MLH), activity by brief description, and numbers of employees.

Geographical coverage: Great Britain

2. Method of Data Collection

The register is updated using information from official returns received by the exchange from some employers (e.g. "L" returns and forms relating to numbers of National Insurance Cards held in June) and from local knowledge, e.g.

(a) the opening or closing of an undertaking which qualifies, or is likely to qualify, as a principal employer, or

(b) any substantial change (say 50% or more, or of over 250 workers, whichever is the smaller) in the number of employees.

3. Geographical Resolution

Basic information unit: Individual employers

Processed data: Employment exchange area

Published data: None

4. Statistical and Legal Restrictions

Statistical limitations: This is not a statistical record, and does not purport to be a comprehensive register of all employers in the area.

Confidentiality: In many cases, the numbers of employees cannot be disclosed. When obtained from returns under the Statistics of Trade Act, they may not be used for any purpose involving breach of the Act. Information may be disclosed to other government departments, but only for statistical purposes not involving breach of the Act.

5. Boundaries

Employment exchange area boundaries.

6. Forms of Data Storage

Original data: Returns from employer and local knowledge.

Processed data: None

7. Operational Details

Originating body: Department of Employment and Productivity

Location of supervisory staff: Local employment exchange

Frequency and dates of survey: The register is subject to annual review, but, if any major changes occur between the annual reviews, the record is revised when they come to notice.

Location of original data: Local employment exchange

Location of processed data: Local employment exchange

Publications: None

1. Subject **DEP Employment Record 2 – Local Employment Records (June employment estimates)**

Content: Topics covered are: numbers of employees, male and female separately, analysed by industry. These estimates are based on the numbers of National Insurance cards exchanged, the quarter commencing on the first Monday in June, supplemented by information supplied by employers.

2. Method of Data Collection

National Insurance cards are exchanged each year, and approximately 25% fall due for exchange each quarter of the year. A count is made of all cards exchanged. Cards exchanged in the June quarter are also classified by industry. In addition, information is obtained from employers of five or more employees, showing:

(a) the number of cards held which are due for exchange in the current quarter;

(b) the number of cards held which will be due for exchange in later quarters; and

(c) the number of cards held centrally for employees who work at other establishments outside the area, and the numbers (in groups of ten or more) working in each area.

This information is linked with the count of cards exchanged to produce estimates for the numbers of employees, male and female separately, for each area analysed by industry. The area is normally that of an employment exchange.

3. Geographical Resolution

Basic information unit: National Insurance cards for employees (subject to confidentiality restrictions — see 4. below).

Processed data: Annual summaries for each employment exchange, of cards exchanged, cards held, and details of cards held for workers in other areas (subject to confidentiality restrictions — see 4. below).

4. Statistical and Legal Restrictions

Statistical limitations: Sampling errors and problems of coverage affect ER 2's considerably. The sampling error for a particular category (e.g. an individual industry in a single employment exchange area) depends on the size of the category and the extent to which information has been obtained from employers in that category. In a cell of less than 10,000, the margins of sampling error may be larger than 2%. Other limitations are:

(a) Some persons are allocated to the area in which their cards are exchanged, if information about the location at which they actually work is not obtainable.

(b) Some classes of workers, e.g. civil servants without National Insurance cards, are not included.

Confidentiality: For users outside government, figures for two or more industries are amalgamated where necessary to ensure that it is not possible to determine the characteristics of an individual establishment. ER 2's for years prior to 1960 cannot normally be made available.

5. Boundaries

Particulars about employment exchange boundaries and group area boundaries are available from DEP.

6. Forms of Data Storage

Forms ER 2 (one sheet per employment exchange or group area) are held at Watford.

Note: Any time series analysis should take account of changes in employment exchange boundaries during the period, and of changes in industrial classification.

7. Operational Details

Originating body: Department of Employment and Productivity (in conjunction with the Department of Health and Social Security).

Location of supervisory staff: DEP,
Orphanage Road,
Watford, Hertfordshire.

Frequency and dates of survey: Employment Records 2 are compiled annually.

Location of original data: Employment exchanges

Location of processed data: DEP, Watford.

1. Subject **Numbers of unemployed persons registered for work at DEP and associated local offices, or Registered Unemployment.**

Content: Local Offices maintain registers in the form of individual records for unemployed persons registered for work. In addition to personal details such as age, the unit includes a record of employment, the industry in which the registrant was last employed, and the occupation for which he is considered industrially best suited.

Geographical coverage: Great Britain.

2. Method of Data Collection

From the information contained in each registrant's records (see 1. above), Local Offices on the statistical date (usually the second Monday in each month), prepare analyses of the unemployed register, by industry, occupation, duration of unemployment, and also showing particular types of registrant such as casual workers. Separate figures are compiled for men, women, boys and girls, and also for the wholly unemployed (including casuals) and for those temporarily stopped. At less frequent intervals, the analyses are more detailed, e.g. for age, duration of unemployment and occupations.

During the four days after the count, adjustments are made for belated information. The completed returns are then forwarded to DEP Headquarters at Watford where they are processed to provide national and regional analyses.

3. Geographical Resolution

Basic information unit: Individual registrant's records (subject to confidentiality restrictions – see 4. below).

Processed data: Local Office by Minimum List Heading of Standard Industrial Classification.

Published data: Local Office totals, or "travel to work area" totals and percentage rates.

4. Statistical and Legal Restrictions

Statistical limitations: The returns cover only those persons who register as unemployed at Local Offices; for this reason the count does not represent total reserves of labour.

Confidentiality: The names and addresses of registered unemployed persons are regarded as confidential. In general, however, there is no confidentiality restriction on the supply of information about aggregate local unemployment.

5. Boundaries

Employment exchange area boundaries.

6. Forms of Data Storage

Original data: The main Local Office returns – e.g. industrial analyses – are kept for about five years and then destroyed, except that returns for one month in each year have been retained since 1948.

Processed data: The punched cards concerned are normally kept for two to three months.

7. Operational Details

Originating body: Department of Employment and Productivity

Location of supervisory staff: DEP, Statistics Division, Orphanage Road, Watford, Hertfordshire.

Frequency and dates of survey: Monthly counts, with more details at less frequent intervals (also see earlier reference to statistical date).

Location of original data: Local Offices

Location of processed data: DEP, Watford

Publications: Extensive and regular statistics are published in the Employment and Productivity Gazette.

1. Subject **Number of Notified Vacancies Remaining Unfilled, or Unfilled vacancies.**

Content: Local Offices (and also Careers Offices) maintain order cards, each of which records the number, and other details of vacancies notified by employers. By inspection, the number of vacancies outstanding or unfilled, at the reference date, can be ascertained.

Geographical coverage: Great Britain

2. Method of Data Collection

From the information contained on each order card (see 1. above), Local Offices or Careers Offices, on the statistical date (usually the first Wednesday in each month) prepare analyses of unfilled vacancies by industry. Separate figures are compiled for men, women, boys and girls. Certain other analyses for men and women provide similar information by occupation. Completed returns are forwarded to DEP Headquarters at Watford where they are processed to provide national and regional analyses.

3. Geographical Resolution

Basic information unit: Individual order card for employer's vacancy. (Subject to confidentiality restrictions, see 4. below).

Processed data: Local Office or Careers Office, by Minimum List Heading of Standard Industrial Classification (or, alternatively, by occupation – men and women only).

Published data: Regional totals: Industrial Groups for Great Britain only (monthly). Occupational groupings for Great Britain, published quarterly. Similar occupational groupings for regions published annually.

4. Statistical and Legal Restrictions

Statistical limitations: The returns cover only vacancies notified to Local Offices and Careers Offices by employers; for this reason, the figures do not represent total demand for labour.

Confidentiality: The names, addresses, and other details of the employer and certain of the information about the vacancy are regarded as confidential. In general, however, there is no other confidentiality restriction on the supply of information about the number of local vacancies remaining unfilled, except where figures for an individual industry or group would reveal the position of an individual employer.

5. Boundaries

Employment Exchange or Careers Office area boundaries.

6. Forms of Data Storage

Original data: The main Local Office returns – e.g. industrial analyses – are kept for about five years and then destroyed, except that returns for two months in each year have been retained.

Processed data: The punched cards concerned are normally kept for one to three months.

7. Operational Details

Originating body: Department of Employment and Productivity

Location of supervisory staff: DEP, Statistics Division, Orphanage Road, Watford, Hertfordshire.

Frequency and dates of survey: Industrial analyses each month; complete occupational analyses at less frequent intervals (also see earlier reference to statistical date).

Location of original data: Local Office or Careers Office

Location of processed data: DEP, Watford

Publications: Extensive and regular statistics are published in the Employment and Productivity Gazette.

1. Subject — **Gross Earnings**

Content: Gross annual earnings of persons in civil employment, classified by age, sex, marital status and area of residence.

Geographical coverage: Great Britain

2. Method of Data Collection

The analysis is obtained from a 0.5% sample of insured persons, i.e. those whose National Insurance number ends in 14A or 84A and whose flat-rate contribution year therefore starts on the first Monday in March, and overlaps the income tax year by some eleven months. Details of gross annual earnings are extracted from microfilms of PAYE documents for men aged 18 to 69 and women aged 18 to 64 in the sample who

(a) had paid at least 48 Class 1 (employed person) contributions in the contribution year; or

(b) had paid at least one, but less than forty-eight, such contributions and had at least forty-eight Class 1 contributions paid or credited (usually for sickness or unemployment) in the year.

Tabulations are produced for group (a) and for groups (a) and (b) together.

3. Geographical Resolution

Basic information unit: Postal address of individual taxpayer (subject to confidentiality restrictions).

Processed data: Counties and large county boroughs (subject to confidentiality restrictions and statistical reliability).

Published material: National and regional levels only.

4. Statistical and Legal Restrictions

Statistical limitations: The results are subject to sampling error, and the extent of this depends upon numbers in the sample and the dispersion of the amounts received in each area. Analyses for married women and widows are less representative than those for men and spinsters, because many employed married women and widows – especially those working part-time – have low earnings which do not bring them within the scope of the income tax PAYE arrangements.

Confidentiality: Individual records are regarded as highly confidential. Aggregations produced by processing, and those used for publication do not involve risk of disclosure.

5. Boundaries

Economic Planning Regions (HMSO).

6. Forms of Data Storage

Original data: PAYE and other records which give National Insurance number, gross annual earnings and postal address.

Processed data: The microfilmed PAYE data are transferred to punched cards and magnetic tape. Locational coding is by area of residence, and is limited to counties and larger county boroughs.

7. Operational Details

Originating body: Department of Health and Social Security

Location of supervisory staff: John Adam Street, London WC 2 and Newcastle-upon-Tyne.

Frequency and dates of survey: Annually since 1964-65.

Location of original data: PAYE and other records at Newcastle-upon-Tyne.

Location of processed data: Newcastle.

Publications: Results of the analysis have been published in the Abstract of Regional Statistics (Vols. 3, 4 and 5 only).

1. Subject **Insured Population**

Content: As well as a record of flat-rate contributions, there is information on age, sex, address, marital status, and industrial category of employment, plus National Insurance Number.

Geographical coverage: Great Britain

2. Method of Data Collection

Records of flat-rate contributions and other associated information are maintained at Newcastle-upon-Tyne on individual ledger sheets. Insured persons are divided into four groups indicated by the suffix A, B, C and D to the National Insurance number, and their contribution cards are exchanged in March, June, September and December. Statistics of the insured population are obtained from a 2% sample, i.e. persons whose National Insurance number ends in 14 or 84, one quarter of the sample being taken from each of the four groups. The sample data which does not include address or other local identification is processed on a System 4 installation at Newcastle. Information for a separate 4% sample of persons in Group B is provided for the Department of Employment and Productivity each year.

3. Geographical Resolution

Basic information unit: Records for individuals' contribution to the scheme including a place of residence address. (Subject to confidentiality restrictions.)

Processed data: Files for individuals. (Subject to confidentiality restrictions.)

Published material: National figures only.

4. Statistical and Legal Restrictions

Confidentiality: Records of individuals are regarded as confidential; aggregated data for sub-areas would not involve disclosure.

5. Boundaries

Not applicable.

6. Forms of Data Storage

Original data: Insurance cards with contribution stamps on them are exchanged annually by employers, or the insured persons themselves.

Processed data: Records of contributions are accumulated at Newcastle on manual record sheets, filed by last two digits of insurance number. Samples of this material are recorded on punch cards.

7. Operational Details

Originating body: Department of Health and Social Security.

Location of supervisory staff: John Adam Street, London WC 2 and Newcastle-upon-Tyne.

Frequency and dates of survey: The record for each person contributing to the scheme is up-dated annually in quarterly groups. The system has been in continuous operation since 1948

Location of original data: Newcastle-on-Tyne

Location of processed data: John Adam Street, London, W.C.2. and Newcastle-on-Tyne

Publications: Annual report of Department of Health and Social Security.

Note on Other Data Items of the Former Ministry of Social Security

Retirement pensions: At present, records of retirement pensions are maintained at Newcastle-upon-Tyne on individual ledger sheets, although the renewal pension order books are normally produced by means of punched cards which show name, pension number, address and post office of payment. The records are being computerised on a System 4 installation. Statistics of retirement pensions are obtained from a 5% sample, i.e. one half of the pensioners whose number ends in the digit 4. The sample data which are processed on the System 4 computer do not include address or other local identification.

Graduated contributions: Graduated contributions are paid with PAYE income tax. The PAYE forms are microfilmed at Newcastle and the information about contributions is later transferred via punched cards to magnetic tape for processing on an EMIDEC computer. Statistics of graduated contributions are obtained from a 2% sample of insured persons, i.e. those whose National Insurance number ends in 14 or 84.

Family allowances: Families with only one child are not eligible for allowances. At present, records of allowances which include the number and ages of the children in the family, address and post office of payment, are maintained at Newcastle on individual ledger sheets. The records will eventually be computerised on the System 4 installation. Statistics of family allowances are obtained from a 4% sample, i.e. persons whose allowance number ends in 17, 37, 67 and 87. The sample data which do not include address or other local identification, are processed by computer annually in full, and four times a year in respect of changes.

Supplementary benefits: Summary returns, i.e. monthly counts of live case papers and records of claims, decisions, home visits, etc. made in local offices, are processed centrally in London. These give information about the number of regular weekly payments, and the work involved in making them. In addition, a 1 in 160 sample of supplementary pension and a 1 in 40 sample of supplementary allowance case papers are examined in February, May, August and

November to determine the various amounts of supplementary benefit in payment, and the average amount of rent allowance. The November sample is extended to include more detailed information about the circumstances and characteristics of supplementary benefits recipients. The sample enquiries are processed by computer at Newcastle-upon-Tyne.

General notes:

At present, many Social Security records are maintained manually, and a number of camparatively small samples is analysed by computer for administrative and statistical purposes. Non-locational reference numbers (e.g. National Insurance number, pension number) are generally used to select the samples. Postal addresses are included in many of the records but the use of these for statistical purposes is restricted by confidentiality rules, non-automation and, in any case, lack of fit between postal areas and local authority boundaries. There will be a large amount of computerisation when the new installation at Newcastle comes fully into use. Theoretically, a great deal of analysis could then be undertaken by computer sorting of postal codings.

1. Subject

SH 3 (Hospital Survey)

Content:

The Department of Health and Social Security delegates considerable responsibility to Regional Hospital Boards (fifteen in England and Wales), and to Boards of Governors of Teaching Hospitals. The Department is therefore concerned with trends in the distribution and usage of hospital facilities as a broad management function. The SH 3 summaries provide basic information to this end, covering numbers of beds, average occupation, numbers discharged or died, length of stay, waiting lists. The above relate to in-patients, but similar information is requested about out-patients. The enquiry covers establishments in the public sector only.

Geographical coverage:

England and Wales

2. Method of Data Collection

SH 3 forms are sent to group Secretaries who notify DHSS of group changes. Some of the information is put onto 80-column punch-cards and analysed by counter-sorter machines. The remaining information is analysed by conventional manual methods from the forms. Regional and national summaries are readily available.

3. Geographical Resolution

Basic information unit: Hospital returns (no geographically determined catchment area – see 5. below).

Processed data:
Published data: Regional and national summaries are available from punched card records of basic units.

4. Statistical and Legal Restrictions

Confidentiality: Individual patient figures would be highly confidential, but summary hospital totals on forms SH 3 are not confidential.

5. Boundaries

Hospitals do not have tightly defined catchment areas either in principle or in practice. Much variety is present according to the different specialities, with dispersion of patients being particularly noticeable in the cases of psychiatry and convalescence. Individual doctors' preferences and the personalities involved carry a great deal of weight.

6. Forms of Data Storage

Original data: 3,500 returns are involved (from 2,500 hospitals and 1,000 clinics).

Processed data: Only a proportion of the data is transferred to punch cards (see 2. above).

7. Operational Details

Originating body: Department of Health and Social Security.

Location of supervisory staff: DHSS,
Russell Square,
London, W.C.1.

Frequency of dates of survey: Annual, with a simplified version at a mid-point between the annual returns. Some information relates to the position as at 31st December, and some covers the whole calendar year.

Location of original data: DHSS, Russell Square.

Location of processed data: DHSS, Russell Square.

Publications: Annual report of the DHSS.

Note on Other Data Items of the Former Ministry of Health

A. Hospital Building

DHSS maintain a list of new developments and alterations begun and completed with a value of over £100,000, i.e. a list of large capital projects. (DHSS distinguish between capital and operating expenditure.)

B. Staff

(i) Hospital service: There is detailed information available about doctors employed by hospital authorities (e.g. place of birth, speciality, etc.). Typically, hospital doctors and some professional and technical staff divide their time between a number of different hospitals and clinics, whereas nursing and ancillary staffs are generally employed at a particular hospital. Many hospital workers are part-time and there are, therefore, two ways of presenting staff figures: (i) total numbers, or (ii) whole-time equivalents. The latter is based on the appropriate working week of the various classes of staff.

A series of annual returns relating to 30th September cover (i) medical staff, (ii) nursing staff, (iii) administrative and clerical staff, (iv) professional and technical staff and (v) ancillary staff, e.g. porters and cooks. In addition the mid-year position (31st March) is available for nursing staff. Returns for the last three years covering all categories of staff are held at DHSS, Russell Square, those for earlier years are kept at Hayes, Middlesex, and are available to DHSS within 24 to 48 hours.

(ii) Executive Council Services: General Medical Services – The councils which in England and Wales number 134, provide the Department each quarter with changes of doctors who join or leave their medical lists. Each year as at 1st October information is provided giving the practice structure, number of patients, areas and type of service each doctor is providing. Various statistical tables are produced on manpower showing sub-divisions by age, practice structure, and services provided both nationally and regionally.

Information is held at DHSS, Russell Square.

General Dental Services – The Councils provide, through the Dental Estimates Board, the Department with statistical information relating to dentists operating within their area. From these return tables are produced showing sub-divisions of dentists by age and services provided both nationally and regionally.

Information is held at DHSS, Russell Square.

(iii) Doctors and dentists practising outside the National Health Service: Are not covered in these returns.

C. Patients and Disease

There is no central knowledge of morbidity in patients under the care of a GP, except in the case of notifiable infectious diseases. In hospitals the SH 3 enquiry (q.v.), split by department and speciality, gives some idea of morbidity. A hospital in-patient enquiry is taken as a 10% sample of the non-psychiatric side (every tenth patient admitted), giving name, address, age, religion, occupation, etc. The hospitals send the half-million forms a year to the GRO for computer analysis, which is carried out on a year-by-year basis. 1966 is the most recently published analysis ("Report on Hospital In-patient Enquiry", Part 1, Tables, published by the Ministry of Health and the GRO) with detail for individual RHB's.

D. Hospital Activity Analysis (HAA)

This covers about 42% of non-psychiatric patients. It is anticipated that expansion group by group will lead to 100% coverage of these patients by 1973.

There is a 100% mental health enquiry, the analysis being carried out centrally on the DHSS computer at Fleetwood (Elliott 4400). This is published under the following title: "Psychiatric Hospitals and Units in England and Wales: Ministry of Health Statistical Report Series No. 4: In-patient Statistics from Mental Health Enquiry for Years 1964, 1965 and 1966".

1. Subject **Housing Statistics**

Content:

Monthly housing progress report (Form P 2(Hsg)) for each local authority in England and Wales gives dwellings started, completed and under construction for the local authority, housing associations and private owners. Houses and flats separated; number of bedrooms in completed dwellings. No statutory obligation for authorities to complete returns; usually about 95% of housing progress is covered by returns received by the 11th working day of the following month, and late returns are included in the succeeding months.

Progress of local authority schemes is monitored and checked by means of scheme cards (giving an identifying site number and characteristics of the scheme including numbers of dwellings of different types and sizes, number of persons to be accommodated, site size, density, facilities provided, etc.) based on Form TC 2 which should be sent by the authority within a month of accepting the tender for the scheme; there are about 5,000 active scheme cards at MHLG. The monthly progress report breaks down local authority housing into these named schemes, but addresses are not necessarily given. For all other housing, totals within the local authority area are given with no further geographical breakdown. Similar information is collected in Scotland for local authority building (Housing Form 1) but totals for private owners are only collected quarterly (Housing Form 121).

Housing built for government departments (including the armed services) is reported separately, and recorded by local authority area.

(This entry refers to the major returns of housing progress. Authorities also return details of slum clearance and closures, other losses from the dwelling stock, improvement grants, loans for house purchase, etc. Finally, MHLG and the Welsh Office have organised house condition surveys in England and Wales which give results at national and large regional levels, and for each of the provincial conurbations.)

Geographical coverage: Great Britain

2. Method of Data Collection

Totals are accumulated manually for regions, sub-regions. May be automated later.

3. Geographical Resolution

All data relate to local authority areas; further geographical sub-division would be possible for local authority schemes by applying detailed local authority knowledge, but no further geographical information is available for other housing.

4. Statistical and Legal Restrictions

Statistical limitations: Delayed returns can have marginal effects on monthly estimates of progress at sub-national levels.

Confidentiality: Local authority totals are published quarterly in "Local Housing Statistics" and the "Housing Return for Scotland". Details of individual local authority schemes are not published, and the breakdowns of dwellings by type and size are not published for sub-regional areas.

5. Boundaries

Local authority areas are defined on a set of maps at 2½" scale, and are kept up to date by MHLG. They are published periodically on a county basis at 1 : 100,000 scale. Small scale boundary maps are also available.

6. Forms of Data Storage

Original data: Monthly P 2 forms, one per authority in England and Wales have been collected since 1952 (they were less detailed in the earlier part of this period).

Processed data: Certain data are extracted, but, in the main, those recorded are processed results for regions. Form TC 2 data on local authorities' scheme tenders will be automated this year, on the HMSO (ICL) computer at Norwich, and it is hoped to automate processing of Form P 2 later so that information for England and Wales should then be more easily accessible.

7. Operational Details

Originating body: Ministry of Housing and Local Government in conjunction with Scottish Development Department, Welsh Office, and local authorities.

Location of supervisory staff: MHLG, Whitehall, London, SW 1.
SDD, 43 Jeffrey Street, Edinburgh 1.

Frequency and dates of survey: Monthly returns.

Location of original data: Local authorities' returns for the current and preceding year are kept at MHLG headquarters and SDD; earlier forms for England and Wales are stored in archives, but Scottish forms are not kept.

Location of processed data: MHLG and SDD.

Publications: "Housing Statistics" gives quarterly summaries for Great Britain (including data from Scottish Development Department and Welsh Office) by economic planning regions (including sub-divisions of the South East EPR). "Local Housing Statistics" and the "Housing Return for Scotland" give quarterly figures for individual local authorities.

1. Subject — **Epitome of Accounts** (Form H31)
Return of Income and Expenditure (Form H 20) (Miscellaneous authority Income and Expenditure)

Content: Income (grants, receipts, transfers) and expenditure (general, principal, interest, sinking fund payments) for about 100 items, separately (as appropriate) for Rate Fund Account, Housing Revenue Account and Trading Services, Special Funds, and Loan and Capital Account. Covers all local authorities (about 1,300 using form H 31) plus miscellaneous authorities (water, river authorities and harbour boards, police forces, etc. but not nationalised undertakings, using Form H 20). Returns are normally received from about 90% of authorities, accounting for about 95% of expenditure.

Geographical coverage: England and Wales

2. Method of Data Collection

Form H 31 (Epitome of Accounts) or Form H 20 (Return of Income and Expenditure) is completed by all authorities and returned to MHLG. Forms from authorities (of a similar kind) are summarised to totals for England and Wales, but not for individual counties. Forms are kept in file for about five years. Older forms are stored in MHLG archives.

3. Geographical Resolution

NB Much of authorities' expenditure is within their own boundaries, but some authorities spend money on housing outside these; furthermore, considerable sums are spent from time to time in, for instance, Wales and the Lake District, by authorities in the Midlands and North West. This fact acts as a general limitation to the geographical resolution of this data.

Basic information unit: All figures represent local authority sub-totals and could not be given a more precise location without considerable extra work.

Processed data: Resolution limited to national level.

Published material: National level only.

4. Statistical and Legal Restrictions

Statistical limitations: (See also 3. above). Epitomes are normally received from 97% to 98% of all local authorities, whereas the overall response rate for both forms (H 31 and H 20) is about 90%. The non-responders among miscellaneous authorities therefore represent a higher proportion, but are normally such small spenders that the effects can be neglected.

Confidentiality: No confidentiality restrictions as published in Local Government Financial Statistics.

5. Boundaries

Local authority areas are defined on a set of maps at 2½" scale, and are kept up to date by MHLG. They are published periodically on a county basis at 1 : 100,000 scale. They are also available at smaller scales. Boundaries of miscellaneous authorities are usually statutorily defined, but may not coincide with local authority boundaries.

6. Forms of Data Storage

Original data: About 2,000 forms H 31 and H 20 each year, each covering a single authority. (Note that the authorities can be nested or overlapping.) Data is given to the nearest £ for numerous categories of expenditure and income.

Processed data: National summaries by kinds of authority. No county summaries.

7. Operational Details

Originating body: Ministry of Housing and Local Government in conjunction with the local authorities.

Location of supervisory staff: MHLG, Caxton House, Tothill Street, London SW 1.

Frequency and dates of survey: Annual – covering each financial year.

Location of original data: See location of supervisory staff.

Location of processed data: See location of supervisory staff.

Publications: Local Government Financial Statistics (MHLG) gives annual summary figures for each service. (Institute of Municipal Treasurers and Accountants also produce financial and rating statistics in a series of publications.)

Note: Since 1st April 1969 the data have no longer been collected by the method described. A new series of forms designed for computer processing and altogether more comprehensive in coverage has been issued. A certain amount of sampling is involved. The final form of published statistics has yet to be determined.

1. Subject — **Statutory Register of Planning Applications**

Content: Registers are required to be kept by local planning authorities. The primary purpose is to enable any person to find out what applications for planning permission have been made and are under consideration, and what decisions have been taken on previous applications, in relation to particular land. Thus each planning application and the decision on it must be entered in the register. Records show date, site, approval or refusal, brief description of development, name of applicant. In addition, details may be given about conditions imposed and reasons for these, and reasons for refusals. (Under Article 2 (i) of the Town and Country Planning General Development (Amendment) Order 1969 (SI 1969 No. 276) the register of planning applications is required to be kept in two parts, Part 1 to contain copies of planning applications and applications for approval of "reserved matters" until they have been "finally disposed of"; and Part 2 which will be the permanent record to contain the information previously specified in Article 14 of the 1963 General Development Order (SI 1963 No. 709).)

Geographical coverage: All local planning authorities in England and Wales.

2. Method of Data Assembly

Planning decisions taken by Councils are deposited in the statutory register of decisions as and when they occur. The register is normally arranged by street, within local authority areas or parishes. Most planning authorities keep the documents for each application (application form, plans, reports to committee, etc.) in a series of files. In one or two cases (i.e. Southwark, East Sussex) computerised records have been set up. The Southwark system uses a street and address code identifier together with a reference to the appropriate 1 : 1,250 map quadrant. The problem of automating the planning applications and decisions record is complicated by several problems:

(a) Identification of sites, especially in rural areas;

(b) Superimposed and over-lapping applications. A property may be the subject of a large number of applications, varying in characteristics and area, and for a complex development. There may be a large number of minor supplementary approvals, e.g. approval to details, approval to extensions and alterations;

(c) In many places, there is no organised system to establish whether or not planning permissions are acted upon. Even periodic land use surveys do not indicate if a building is old or new. However, new regulations under the Town and Country Planning Act, 1968, impose time limits on the duration of planning permissions, and provide for a procedure to terminate permissions where development has started but has not been completed. Authorities may now have more reason, therefore, to establish whether permissions have been implemented.

3. Geographical Resolution

Planning decisions give address of properties concerned, although it is not always easy to identify the precise boundaries.

4. Statistical and Legal Restrictions

Confidentiality: The statutory register is a public document; internal files are normally confidential.

5. Boundaries

Records relate to individual planning decisions, and the registers are kept in the offices of local authorities (see 3. above).

6. Forms of Data Storage

Original data: Each planning application produces an internal file of documents and a formal decision by the Council.

Processed data: For content of register, see 1. above.

7. Operational Details

Originating body: Local planning authorities.

Location of supervisory staff: Local planning authority offices.

Frequency and dates of survey: Cumulative record as applications are lodged and decisions are taken.

Location of original data/Location of processed data: The register is required to be kept by local planning authorities, that is to say, by County Councils, County Borough Councils and the Lake District and Peak Park Planning Boards, but parts are kept at the offices of the County District Councils, and are likely to be compiled by officers of these Councils.

Publications: MHLG publishes annual tables, related to classes of development, giving numbers of applications received, granted and refused by local planning authorities, and the numbers of appeals lodged, and their results.

1. Subject **Office Development Permits**

Content:

This provides information about permits issued, refused, the gross area (floor space in square feet) covered by permits, and the floor space relinquished by applicants.

Up to July 1967, all office projects above 3,000 square feet were subject to the control. From that date onwards (excepting the Metropolitan region, where it remained at 3,000 square feet), the exemption limit was raised to 10,000 square feet. From February 1969 the exemption limit is 3,000 square feet within the Greater London Council area, and 10,000 square feet elsewhere where the control operates.

Geographical coverage:

The Control Regions: these have changed over time:

(a) From August 1965 onwards, the London Metropolitan Region and the West Midlands conurbation (with retrospective application to November 1964 in the case of the Metropolitan Region).

(b) From July 1966 onwards, the whole of the South East, East Anglia, West Midlands and East Midlands Economic Planning Regions.

(c) From February 1969 onwards, the whole of the South East Economic Planning Region, the West Midland Economic Planning Region except Shropshire and Herefordshire, and the East Midlands Economic Planning Region except Rutland and Kesteven and Holland in Lincolnshire.

2. Method of Data Collection

Under the Control of Office and Industrial Development Act, 1965, planning applications for office development in excess of the exemption limit within the controlled areas had to be accompanied by ODP's from the Board of Trade. The Ministry of Housing and Local Government assumed responsibility from 23rd October 1969. Application forms are sent to the appropriate Regional Office of the Ministry, which may grant or refuse permits for the proposed developments.

3. Geographical Resolution

Basic information unit: Local authorities and DEP employment exchanges.

Processed data: All data processing is carried out manually.

Published material: Economic Planning Regions and certain sub-divisions.

4. Statistical and Legal Restrictions

Confidentiality: To meet the requirements of the Statistics of Trade Act, disclosure is restricted so that details about individual applications are not released.

5. Boundaries

Economic Planning Regions and some sub-divisions (HMSO).

6. Forms of Data Storage

Original data: Application forms do not always specify an actual site, and so the details given about location vary. The precise location is, of course, known before a permit can be issued.

Processed data: Because of the limited numbers of applications for permits under this control, all processing is carried out manually. The size of the job makes it unlikely that any other method will be used.

7. Operational Details

Originating body: MHLG

Location of supervisory staff: MHLG, Economics and Statistics Division,
Whitehall,
London, SW 1.

Frequency and dates of survey: These data are the result of a continuous process of control. However, published data relate to quarters of each year.

Location of original data: Application forms are retained at the regional offices of the Ministry.

Location of processed data: MHLG, Economics and Statistics Division,
Whitehall,
London, SW 1.

Publications: Quarterly data about permits issued, together with a size and regional analysis, are published regularly in the Board of Trade Journal. Data about permits refused are published annually in the Annual Report under the 1965 Act.

1. Subject — **Population Census 1951, England and Wales**

Content:

Final schedule had 24 questions, several of which applied to the household rather than to each individual. Many of the questions were essentially similar to those asked in previous censuses, with new information being sought about household amenities and greater detail on "part-time" workers. Schedule covers: name, relationship to head of household; usual residence; sex; age; marital condition (single/married/widowed – date of marriage, children born in marriage, both in total and a separate note as to whether any were born in the preceding 12 months); birthplace; nationality; education (whether in full or part-time education or, if education complete, date when full-time education finished); occupation and industry (personal occupation, employer and employer's business, place of work); household arrangements for water supply, cooking, etc. (piped water supply, cooking stove or range, kitchen sink, w.c., fixed bath – whether use is exclusive to household, shared or none).

The different topics covered by the Census are analysed and published in a series of summary volumes, described in 7. below.

Geographical coverage:

England and Wales, but censuses were also taken on the same date by the appropriate authorities in Scotland, Northern Ireland, Isle of Man, Channel Islands and Irish Republic.

2. Method of Data Collection

Questionnaires are distributed and collected by enumerators, and provide data for Census night. All information punched onto 65-column cards and sorted by High Speed sorter and UPCS (Universal Printing Counting Sorters). Processing times: sorter – 47,600 hours, UPCS – 36,500 hours.
The cost of the census to the GRO was £1,263,362.

There exists power under Section 4 (2) of the Census Act, 1920, to satisfy on repayment of cost any reasonable demand for statistical information from the Census not covered by the standard programme. (Local authorities, universities and individuals have made use of this.)

3. Geographical Resolution

Basic information unit:

Schedules for individual households – subject to confidentiality restrictions (see 4. below).

Processed data:

None, other than that published. Punched cards were destroyed after processing had been completed. Household data on cards – subject to confidentiality restrictions (see 4. below).

Published material: Ward and parish summaries.

4. Statistical and Legal Restrictions

Statistical limitations: The General Report of the 1951 Census includes a chapter on the quality of the data, which discusses completeness of cover, quality of response, comparability with the 1931 census, and an approval of the preliminary 1% Sample Tables.

Confidentiality: Individual household returns are confidential, and all data made available are in aggregated form, either by area, group or topic. Incidental disclosures occur when there are only one or two households in a depopulated rural parish.

5. Boundaries

Boundaries of ED's and wards are described only verbally. Local authority boundaries are marked on a series of county maps, mostly at 1 : 100,000 scale, which are kept up to date by the Ministry of Housing and Local Government.

6. Forms of Data Storage

Original data: All data come in on the census schedule (one per household) which are delivered and collected by enumerators. Questionnaires include house address; the enumerator adds the following to the form:-

(i) Number of registration district;
(ii) Number of registration sub-district;
(iii) Number of enumeration district.

Processed data: Data on cards: household number, not the address, is transferred to cards. Address can be identified from the household and enumeration district code number by means of the Enumerator's Record Book (held at Titchfield).

7. Operational Details

Originating body: General Register Office,
Somerset House,
London, WC 2.

Location of supervisory staff: Somerset House and GRO,
Census Branch,
Titchfield,
Fareham, Hampshire.

Frequency and dates of survey: Basically decennially with a gap, due to war, in 1941 and, because 1951 was the first census for twenty years, it was decided to publish advance tables on a 1% sample of the returns. The 1951 census was taken at midnight on Sunday, 8th April.

Location of original data: Public Record Office,
Bourne Avenue,
Hayes, Middlesex.

Location of processed data: Titchfield, Hampshire.

Publications:

(a) Preliminary Report – Provisional figures by sex of the population in England and Wales as a whole, by administrative counties, county boroughs, municipal boroughs, urban districts and rural districts (e.g. Loddon RD, Cromer UD). Tables give area and population for 1931 and 1951 by sex, and 1931/1951 change in numbers and as a percentage. Report published three months after Census Day. Derived from local summaries rapidly compiled and then assembled centrally.

(b) 1% sample tables – Drawn from returns for early processing. Results published in two volumes, fifteen months and nineteen months after Census Day. Twelve sections of subject matter. Sample drawn after information filled in on returns and therefore more efficient than if sample had been set up before.

(c) County reports, e.g. Norfolk – Give final population figures and details of dwellings, households, "institutions", by sex, age, marital condition, birthplace and nationality, education and social class. Local figures for occupation, industry, usual residence and workplace not included in county reports – they appear in relevant subject volumes. Population and acreage of parliamentary, local government and petty sessional areas. Information goes down to give figures for urban districts, rural districts, wards and civil parishes (data are given for civil parishes if they have (i) 1,000 or more population, and (ii) a density of 0.3 or more persons per acre).

(d) The Report on the Welsh-speaking population – Statistics about Welsh spoken in Wales (including Monmouthshire).

(e) The Population of Ecclesiastical Areas (England) – Gives population for ecclesiastical parishes and dioceses.

(f) Index of Place Names – Location and enumerated population of every area mentioned in the various census volumes.

(g) Report on Greater London and Five Other Conurbations – Similar in content to the county reports (see (c) above).

(h) Report on Usual Residence and Workplace – Tables of population resident in one area and working in another (boroughs, urban and rural districts). "Tables comparing the enumerated census population in those units, with an alternative allocation based on the statement of usual residence on the Census Schedules, together with the numbers enumerated there who were usually resident outside England and Wales, and tables classifying the latter by country and usual residence and by age". Table 5 is the chief one – of people resident in one area and working in another: the analysis is done both ways: (i) Resident in this area but working elsewhere, and (ii) Working in this area but resident elsewhere. (Figures are given, even if there is only one person in a cell.)

(i) The Housing Report – Summarises information given in the county reports.

(j) The Occupation Tables – Statistics of the occupied population according to their gainful occupation and according to the area in which they are enumerated by age, sex, marital condition, industrial status in employment, social class and socio-economic group. (Metropolitan Boroughs and urban areas whose population exceeds 50,000 are listed.)

(k) The General Tables – Summaries of information given in the county reports on local populations (plus figures for counties and county court districts); "non-private" population by birthplace, nationality, age, marital conditions and education.

(l) Industry Tables – Statistics of the occupied population based on industrial activity to which their occupation contributes by industrial status; married women. Areas (metropolitan boroughs, UD's and RD's) are those of place of work as distinct from place of enumeration or residence. Also analyses important industries (or groups thereof) by main occupations, Minimum List Heading of Standard Industrial Classification being used.

(m) Fertility Report – Statistics derived from questions put to married women less than fifty years in age, on date of marriage and number of children. Some tables based on one-fifth and others on four-fifths sample of available records. Some analysis by regions, conurbations in general and urban/rural aggregates. Analysis, too, by social class, socio-economic groups, industry groups. In other tables, the areal break-down is conurbations, urban areas (outside conurbations) with populations greater than 100,000, urban areas (outside conurbations) with populations between 50,000 and 100,000, and urban areas (outside conurbations) with populations less than 50,000, and rural districts outside conurbations.

(n) General Report – Written at the end of the census analysis and production, giving general description of operation and some general tables.

NB In planning for the 1971 Census, GRO are investigating the possibilities of co-ordinate references for individual households, and of analysis by a nesting hierarchy of grid squares. There is no suggestion of applying any such techniques, still experimental, to historical data.

1. Subject **Census 1961, England and Wales**

Content: Census conducted at two levels:

(a) Full count for some information; and

(b) 10% samples for more detailed information. The full list of questions for 1961 was slightly larger than in 1951 but, because sampling methods were introduced at the enumeration stage, nine out of ten people were asked for less information than at any previous census this century.

The questions asked followed much the same pattern as in previous years: birthplace was simpler than in 1951, only asked for country within the UK, not actual county; nationality section was extended to cover the different Commonwealth citizenships; education questions extended to cover scientific and technological qualifications; a new question was introduced aimed at ascertaining the frequency, amount, direction and characteristics of population movement; housing was also wider than previously and included details of housing tenure. (Schedule E 90, designed for 90% of the people in England, covered: name, relationship to head of household, usual address, sex, age, marital status, number of children, date of marriage, country of birth, place of birth, if outside UK, plus household questions, e.g. tenure, amenities and numbers of people and rooms.)

The 10% sample was introduced at the enumerating level and, hence, there were two different types of forms covering the full population (e.g. E 90), and the 10% sample (e.g. E 10). (Actually there were seven different forms to take account of Wales, institutions, etc.) The sample was considered adequate for some purposes, e.g. analysis of working population. Used, in fact, for occupation, employment, place of work, status in employment, education, scientific and technological qualifications, change of usual residence or duration of stay at present usual residence and also people usually in the private household, but temporarily absent on census night. (Schedule E 10, designed for 10% of the people in England, covered: the same subjects as in E 90 above, plus usual address, and whether any change in the last twelve months; date of end of full-time education (for those over fifteen); science and technology qualifications; name and business of employer, occupation, place of work, whether full or part-time, and, for men (if part-time), details of last full-time employment; if retired or sick, details of last employment. And details of those absent from the household: name, relation to head of household, sex, age, marital status, place of abode on Census night, name and business of employer, occupation and, if retired, sick or out of work, details of last employment.)

The different topics covered by the Census are analysed and published in a series of summary volumes, described in 7. below.

Geographical coverage: England and Wales, but censuses were also taken on the same date by the appropriate authorities in Scotland, Northern Ireland, Isle of Man and the Channel Islands. Census taken in the Irish Republic on 9th April 1961.

2. Method of Data Collection

Local enumeration is more or less based upon the areas covered by local registrars of births and deaths. The distribution and collection of the schedules were handled as previously, by a small army of enumerators, and it was at the enumeration level that the 10% sample was introduced. Three important innovations were introduced in the 1961 Census:

(i) Extra questions were asked of a one in ten sample of the population;

(ii) Computer processing was introduced;

(iii) Automatic typing of the results and photographic reproduction (the computer output information to punched cards which were then fed to a "card-a-type" machine).

No 1% sample preliminary tables were planned, as had been undertaken in 1951, because the GRO hoped to publish all the 1961 tables in two or three years. In the event, unforseen difficulties gave rise to considerable delays. 15,000 hours of computer time were used for processing, and the cost to the GRO was about £2,697,000. The 10% sample turned out to be biased, and therefore suitable adjustments had to be calculated.

After much discussion about which would be suitable computers, the GRO decided to use the IBM 705 machine installed for the Royal Army Pay Corps at Worthy Down, Winchester.

A post-enumeration survey was conducted to assess the completeness of enumeration and the quality of answers given, i.e. this provides a check on coverage and quality.

3. Geographical Resolution

Basic information unit: Schedule E 90 (full census) — Schedules for individual households subject to confidentiality restrictions (see 4. below).

Schedule E 10 (10% sample for supplementary information) – Schedules for individual households – subject to confidentiality restrictions (see 4. below). Geographical resolution restricted by level of sample reliability.

Processed data: Household data on tape – subject to confidentiality restrictions (see 4. below). Ward and parish summaries.

Published material: Ward and parish summaries.

4. Statistical and Legal Restrictions

Statistical limitations: For the 1961 Census, a statistical assessment programme was carried out which was much more extensive than in any previous census. A very full account of this appears in the General Report, and discusses completeness of coverage, sampling errors, quality of response and the statistical impact of processing procedures. There is also a chapter on comparability with the previous (1951) Census.

Confidentiality: Individual household returns are confidential, and all data made available are in aggregated form, either by area, group, or topic. Incidental disclosure occurs where there are only one or two households in a depopulated rural parish.

5. Boundaries

Boundaries of ED's and wards were only described verbally. Local authority boundaries are marked on a series of county maps, mostly at 1 : 100,000 scale, which are kept up to date by the Ministry of Housing and Local Government.

6. Forms of Data Storage

Original data: As in 1951, raw data are on census schedules, delivered and collected by enumerators. The forms have the household address on them. There are about sixteen million schedules.

Processed data: Most data from schedules are transferred to magnetic tape: individual records are identified by household code number (tallying with names and addresses in the enumerators' Record Book kept at Titchfield). Ward library summary tapes (including enumeration district summaries) are available for certain areas. 10% sample data are not entirely reliable at an enumeration district level.

7. Operational Details

Originating body: General Register Office, Somerset House, London, WC 2.

Location of supervisory staff: Somerset House, and GRO, Census Branch, Titchfield, Fareham, Hampshire.

Frequency and dates of survey: See notes for 1951 Census. The 1961 Census relates to midnight of Sunday, 23rd April.

Location of original data: Schedules stored at: Public Records Office, Bourne Avenue, Hayes, Middlesex.

Location of processed data: Data tapes kept at Titchfield, Hampshire.

Publications:

(a) Preliminary Report – Published in June 1961. General notes on census-taking and statistical approach. Also provisional population figures based on summaries of the enumerators (detail for urban districts, and rural districts).

(b) County reports, e.g. Norfolk – Statistics of local interest collected on 100% basis. Covers population and acreages of local government areas; distribution of local population by sex, age, marital conditions, birthplace, citizenship and nationality, private households and size distribution, buildings, dwellings and rooms in which they live; tenure of accomodation and certain household arrangements. Information for MB's, UD's and RD's with ward and civil parish detail for some tables. These county reports contain no information collected on a sample basis.

(c) Report on Scientific and Technological Qualifications – 10% sample information. Statistics of persons thus qualified with details of occupation and industries in which they work.

(d) Report on the Welsh-speaking Population – Relates to Wales and Monmouthshire as for 1951 Census. Published 1962.

(e) Usual residence tables – Published in October 1964. They compare the enumerated census population in local areas with the population usually resident there. Also those enumerated in an area but usually resident outside England and Wales. Detail is given for MB's, UD's and RD's. (Note that this volume of tables is quite separate from the work-place tables.)

(f) Age, Marital Conditions and General – Published in November 1965. Summaries of county report information.

(g) Birthplace and Nationality – Summaries of county report information.

(h) Household and Household Composition – Summaries of county report information and more extensive tabulations for England and Wales.

(i) Migration Tables – Based on 10% sample information. Numbers and characteristics of people who changed usual residence in the year before Census Day. Details of age, marital condition, socio-economic group, occupation and industry, and information about "the rest" classified by length of time at present address. Information goes down to MB, RD, and UD if there are ten or more emigrants or immigrants.

(j) Workplace Tables – 10% sample information. Show the population resident in one place and working in another by MB's, UD's, RD's and New Towns. Also gives comparisons of resident and day populations. The analysis is done by both emigrants and immigrants to a local authority area.

(k) Occupation Tables – 10% sample information. Classified by "1960 Classification of Occupation". Analysed by age, marital condition, employment status and socio-economic group. Most of this information is given on a national basis, but some analysis is done by area types (e.g. conurbations, rural areas).

(l) County Tables – Small additional tables giving information about occupation, industry and socio-economic groups, e.g. Norfolk (administrative county with county boroughs), Great Yarmouth CB, Norwich CB and the administrative county. All the information in these tables is based on sample data, and they therefore differ from the main county volumes.

(m) Industry Tables – Identify employment status, age-groups, married women, SIC classifications and the branch of economic activity to which they contribute. Local population statistics are based on the area containing the place of work. There is some analysis of each important industry or group of industries by the occupations it contained. The tables are based on 10% sample information, and this is given for regions and conurbations. More detailed information is presented in the various county leaflets.

(n) Education Tables – The figures relate to terminal education age of the people in local authorities. National and regional tables give this and age and occupation.

(o) Fertility Tables – These statistics are derived from questions asked of any woman who had ever been married, about the date of marriage and number of children. Nearly all national figures, except for one important table giving information about regions, conurbations and urban/rural aggregates.

(p) Commonwealth Immigrants Tables — Information about immigrants in conurbations.

(q) Greater London Tables — Published after the establishment of the Greater London Council.

(r) Great Britain Tables — Population, birthplace and nationality, housing, household composition, migration, occupation, industry, education and fertility for Great Britain as a whole.

(s) Index of Place Names — Location and census population of places with defined boundaries.

(t) County Leaflets — Population by sex and total number of households and dwellings. Published in advance of the main series of county reports.

(u) National Leaflets — Early information on national and regional basis for household composition, migration, occupation and industry.

(v) General Report — Published in 1968. As in 1951, this gives a general description of the operation.

NB In planning for the 1971 Census, GRO are investigating the possibilities of co-ordinate references for individual household, and of analysis by a nesting hierarchy of grid squares. There is no suggestion of applying any such techniques, still experimental, to historical data.

1. Subject **Sample Census 1966, England and Wales**

Contents: The 1966 Census was not complete, but was based upon 10% of the population, selected thus:

(a) 1 in 10 of the structurally separate private dwellings and "small" non-private establishments.

(b) 1 person in 10 at "large" non-private establishments. A computer print-out listed those to be sampled (see 2. below for detail).

The schedule covers name, relationship to the head of the household, whether present for the night or arriving in the early morning, usual residence, usual address one year ago and five years ago, sex, birthdate, marital status, birthplace, employer, occupation (full-time or part-time), place of work, method of transport to work, unemployed (reason, last employer, last occupation), degree, etc., rooms, ownership and renting, household and amentities, cars and garaging.

Geographical coverage: England and Wales (GRO Edinburgh, responsible for the Census in Scotland).

2. Method of Data Collection

The sample frame for the 1966 Census was the 1961 Census record; the sample was stratified by local authority area. The presentation convention of placing a comma between the second and third digit from the right in entries with three or more digits has been observed to emphasise that the figures shown have been derived from a 10% sample and have not been multiplied by 10. Post-enumeration surveys were conducted to check the results. As this was solely a sample enquiry, the enumeration districts used are considerably larger than those used in 1961. Those used in 1971 will have smaller populations than those for 1961, as extra enumeration work will be entailed.

Most of the information on the schedules is transferred to magnetic tapes, including a household code number. Addresses can be traced from the household code number via the enumerators' Record Books kept at Titchfield. Ward library summary tapes (including enumeration district summaries) have been produced for all areas although in some parishes the sample is very small. Processing was on a similar IBM 705 computer used in 1961, which has been moved to Titchfield. An ICL 1904 machine has been installed at Titchfield for use on the 1971 Census, which is likely to involve nearly twenty million household schedules.

3. Geographical Resolution

Basic information unit: Schedules for individual households — subject to confidentiality restrictions (see 4. below). Geographical resolution restricted by level of sample reliability.

Processed data: Household data on tape — subject to confidentiality restrictions (see 4. below). Geographical resolution restricted by level of sample reliability. Ward and enumeration district summaries.

Published material: Local authorities.

4. Statistical and Legal Restrictions

Statistical limitations: In England and Wales the 1966 Census was based entirely on a 10% sample enumeration, stratified by local authority area. Each published report gives a general indication of the sampling errors involved in the results, but more detailed calculations will be published as part of the final statistical assessment of the Census.

Confidentiality: Individual household returns are confidential, and all data made available are in aggregated form, either by area, group, or topic. Incidental disclosure occurs where there are only one or two households in a depopulated rural parish.

5. Boundaries

In this Census, ED and ward boundaries were for the first time marked on OS 6" maps. Local authority boundaries are marked on a series of county maps, mostly at 1 : 100,000 scale, which are kept up to date by the Ministry of Housing and Local Government.

6. Forms of Data Storage

Original data: Hand written schedules, each containing household names and addresses.

Processed data: Raw data and ward library summary tapes.

7. Operational Details

Originating body: General Register Office,
Somerset House,
London, WC 2.

Location of supervisory staff: Somerset House, and
GRO, Census Branch,
Titchfield,
Fareham, Hampshire.

Frequency and dates of survey: See notes for 1951 Census, Sample Census 1966 relates to midnight of 24th April 1966.

Location of original data: Public Records Office,
Bourne Avenue,
Hayes, Middlesex.

Location of processed data: GRO, Titchfield, Hampshire.

Publications:

(a) County reports – Statistics for administrative counties, county boroughs, local authorities with more than 15,000 population, and county aggregates covering population and acreage; sex, age and marital condition; birthplace; private households, dwellings, tenure and household amenities; private motor car and garaging; economically active and retired males and their socio-economic groups.

(b) Great Britain Summary Tables – Summarising information given in county volumes.

(c) Economic Activity Tables – To be published in four volumes. To be based, as formerly, on (i) occupation and (ii) industry, by age, marital condition, employment status, socio-economic group. Analyses of major industries by contributory occupations. County leaflets will cover counties and large local authorities in England, Wales and Scotland.

(d) Migration Tables – Two main volumes for England and Wales, and nine regional volumes (e.g. East Anglia). Tables 1 and 2 give numbers and characteristics of people who changed their usual residence in the year before or in the five years before Census Day, with details of sex, age, marital condition, and also the number and characteristics of migrant households and families. The regional reports will repeat some Part 1 information, and, in addition, will give occupation, industry, socio-economic group and area of previous residence.

(e) Workplace/Transport Tables – Two volumes for England and Wales and one for Scotland giving details of resident population and population employed in an area, together with the main workplace movements, main means of transport for such movements, and characteristics of people in particular streams. Detail given for urban districts and rural districts. Tables differ from 1961 in that, in 1966, the information is split into two separate tables for residence and workplace.

(f) Household Composition – Composition and social and economic characteristics of households and families.

(g) Usual Residence and Birthplace – Will compare enumerated census population in local areas with population usually resident in those areas and those enumerated who are usually resident outside England and Wales by country of usual residence, marital conditions and age.

(h) Housing Tables – Two volumes for England and Wales. Will cover things not in GB summary tables – dwellings, households, tenure, household amenities, cars and garaging. Some local authority tables. Statistics also for non-private establishments.

(i) Parliamentary Constituency Tables — Sex, age marital condition, economic activity, dwelling, household, tenure, amenities, cars and transport to work by constituency areas of the United Kingdom.

(j) Commonwealth Immigrant — Distribution and characteristics of people born in Commonwealth (except Canada, Australia and New Zealand).

(k) Education Tables — Age, sex, occupation and industry for those who had obtained qualifications (academic or vocational) higher than A-level since the age of eighteen, and details of type, level and subject of qualification.

(l) Report on Special Study Area — Will relate to that part of Scotland which was enumerated on 100% basis, and will be comparable with the county reports. Figures will be given on the fullest scale, irrespective of the size of the local authority.

(m) Commentaries — Not in any census volumes except for the Special Study Area but may be published at some later stage.

(n) Administrative Report — Will appear after all the other volumes have been produced.

(o) Statistical Report on Census Methodology — Will appear after all the other volumes have been produced.

NB In planning for the 1971 Census, GRO are investigating the possibilities of co-ordinate references for individual households, and of analysis by a nesting hierarchy of grid squares. There is no suggestion of applying any such techniques, still experimental, to historical data.

1. Subject **Census of Production**

Content:

The Census covers, among other things, inputs and outputs, capital investment and employment for manufacturing industry, mining, construction and power production, both public and private. (SIC Orders 2 - 18.) It does not cover retailing, wholesaling, the services or transportation; also, where possible, all "non-productive" activities are excluded, e.g. warehousing, offices. However, where for accounting reasons such exclusion is impossible, the establishment is defined as including the activity.

There is a standard questionnaire form sent to smaller firms (less than twenty-five employees) which only covers type of work and numbers employed. A "Long Form", which is sent to larger firms, includes questions on employment, wages, stocks, capital expenditure, work given out, payments for repairs, maintainance, transport rates, plant hire, post, telephones, materials, fuel and power, and payment received for sales and work done. Finally a shortened version of the "Long Form" is also sent to a sample of small firms where these are important to a particular industry.

Employment is defined as the average of the figures for the last week of each calendar month during the year in question.

Where possible, information is gathered for individual establishments, but combined returns (covering more than one establishment) are accepted where it is difficult for a respondent to provide information in greater detail, provided that the establishments included in the return are all in the same industry and in the same country (England, Scotland or Wales). In the 1968 Census a special effort was made to persuade firms to provide information in greater detail.

Geographical coverage: Great Britain (Censuses of Production for Northern Ireland are taken by the Northern Ireland Government under the Statistics of Trade Act (Northern Ireland) 1949).

2. Method of Data Collection

(a) Full census for 1963; postal questionnaires mailed to all firms in the industries listed in Orders 2 - 18 of the Standard Industrial Classification, i.e. manufacturing industries, mines, quarries, construction and gas; register based on that used for previous census, plus changes noted by the Register of Companies, plus notifications from the then Ministry of Labour, factory inspectors, and various other sources. Census was for calendar year 1963, or for business year ending between 6/4/63 and 5/4/64.

(b) Simplified sample enquiry for each year; typically covers only value of capital expenditure, stocks, and work in progress. Made up of about 2,000 large firms who make voluntary quarterly or monthly returns to the Board of Trade, 1,600 other large firms selected and approached, and a sample of 3,500 firms employing more than twenty-five persons in the construction industry. (That is, emphasis is on getting approximation of volume of investment activity rather than production or employment.) It is important to note that the basic reporting unit is not the establishment, as in the case of the quinquennial Censuses, but the business or management unit, which is a broader concept than that of the establishment.

(c) For the future, it is intended to replace the existing system (including the quinquennial Censuses) with a new system of industrial statistics based on a Central Register of Businesses. Under the new system there will be annual Censuses of Production covering larger firms (normally those with twenty-five or more employees, but the cut-off will be lower in industries where smaller firms are important); quarterly sales enquiries with the same coverage; and triennial enquiries into purchases – as well as other, infrequent, enquiries into specific topics. (Further details of these developments may be found in the May, 1968 and the February, 1969 issues of the Statistical News.)

3. Geographical Resolution

Basic information unit:

1958 – Questionnaires for individual establishments (subject to confidentiality restrictions – see 4. below).

1963 and 1968 – Questionnaires for individual production establishments, but some returns are combined. (Subject to confidentiality restrictions – see 4. below.) The questionnaires include the postal address.

1970 and onwards – The new system of industrial statistics will, for the time being, use the employment exchange area as a locational code. Eventually either the national grid or a co-ordinate referencing system, or both, may be incorporated.

Processed data:

1958 and 1963 – Individual returns (subject to confidentiality restrictions – see 4. below). Summaries by "district", an aggregation of Department of Employment and Productivity employment exchange areas. Each return has a "district" reference number.

1968 – Individual returns – will be coded by employment exchange. (Subject to confidentiality restrictions – see 4. below.)

Published material:

1958 – Figures for Census of industries by Standard Region for selected indicators (number of establishments, total sales, employment).

1963 – Regional analyses have been published (1970).

4. Statistical and Legal Restrictions

Statistical limitations:

The increasing use of sampling techniques, particularly for the smaller firms involved, adds the problem of sample reliability to that of non-response. In the compilation of published reports, estimates are made in order to take these deficiencies into account, and to obtain totals for industries as a whole.

The acceptance of combined returns also involves estimation for the purpose of any detailed geographical analysis or any analysis by size of establishment. For these purposes, the figures returned are apportioned to the individual establishments in proportion to the number of operatives.

Confidentiality:

Disclosure is restricted by the Statistics of Trade Act, 1947. This Act specifically prohibits the publication of the number of returns received with respect to the production of any article if that number is less than five. The Board of Trade is further enjoined by the Act so to arrange reports that no particulars can be identified as relating to any individual undertaking, unless the person carrying on the undertaking gives his consent in advance. Finally, the Board may impose further restrictions if they are thought desirable.

As a result, where there are less than five entries contributing to a commodity heading, the value is shown but the number of entries is suppressed. Moreover, in practice, whatever the number of entries for any given commodity total, if one or two firms account for a very high proportion of that total, the figures are only published with the prior consent of the firm in question. If that consent is not obtained, the figures are merged with another heading. Separate reports for industry sub-divisions are not published where less than three enterprises have reported (except in the case of certain nationalised undertakings).

Under the terms of the Act, returns may be disclosed in accordance with directions from the President of the Board of Trade to a Government Department, or they may be disclosed for the purposes of legal proceedings. Otherwise, no information relating to individual undertakings may be disclosed without the prior consent of the person carrying on the undertaking.

5. Boundaries

Employment districts are amalgamations of DEP employment exchange areas; boundaries may be obtained from the Department of Employment and Productivity or the Board of Trade. 1958 Census involved Standard Regions (HMSO).

6. Forms of Data Storage

Original data: Original data are on returned questionnaire forms. In the quinquennial Census, about 130,000 establishments are covered, though there are slightly fewer returns because of combined replies. Each questionnaire has a postal address.

Processed data: Surveys prior to, and including 1958, used punched cards. Processed data from the 1963 and 1968 Censuses are stored on magnetic tapes for use on LEO 3 computers. Summaries are available by industry, but are also available by employment districts (amalgamations of employment exchange areas). Printouts show line summary for each return, and are arranged by industry and size of establishment. The 1968 Census will be coded by employment exchanges. Tapes are assembled as returns are recieved.

7. Operational Details

Originating body: Board of Trade

Location of supervisory staff: Business Statistics Office,
Lime Grove, Eastcote,
Ruislip, Middlesex.

Frequency and dates of survey: Full Census: 1948, 1951, 1954, 1958, 1963, 1968.
Simplified enquiry: All other years – the use of samples began in 1952.
It is hoped to be on a continuous reporting basis by 1970.

Location of original data: See location of supervisory staff.

Location of processed data: See location of supervisory staff.

Publications: Published material is grossed up to cover non-response, and gives summaries for each of a large number of industrial categories for the country as a whole. From the 1958 Census figures, figures have been produced for Census industries by Standard Region; regional analyses of the 1963 Census have been published (1970).

1. Subject **Census of Distribution**

Content: The Census of Distribution records identity, address, kind of business, form of organisation, employment, wages, turnover, purchases, stocks, gross margin, instalment credit, commodity analysis of turnover, and capital expenditure in retail trade. Generally excludes services, but covers establishments for hiring or repairing consumer goods, boot and shoe repairs, laundries, and hairdressing. It excludes catering, wholesale and motor trade, builders' merchants, street sellers and auction rooms (except in 1950 when coverage was much wider, but the range of questions narrower). Data are recorded for establishments (a separate place of business or retailing point within the scope of the Census) and for organisations. Kind of business, form of organisation, turnover and employment are given for both, the remaining data listed above are normally given for organisations. Turnover is the total trading receipts, including meals and services, purchase tax, delivery, and HP charges, less discount and credit. Employment is all persons who worked in the establishment during a specified week, even if temporarily absent, or not receiving a formal wage. Part time workers are included and identified.

Geographical coverage: Great Britain

2. Method of Data Collection

(i) Full census for 1961; conducted by post during 1962. Information was required under Statistics of Trade Act, 1947, for calendar year 1961 or business year ending between 6/4/61 and 5/4/62. The register used for mailing questionnaires was based on that used for earlier censuses (for larger traders), supplemented by a visual enumeration by the GRO and local authority returns (for smaller traders, market and street vendors). Employment information was required for the week ending 28/10/61. A sample follow-up of non-respondents was carried out in September/October 1962. Satisfactory response was obtained from 88% of establishments, representing 93% of estimated total turnover.

(ii) Sample census for 1966; this consisted of (a) full coverage of all co-operatives, organisations with ten or more establishments, mail order firms, and large establishments of independent traders (these were defined as establishments with a turnover in 1961 exceeding £50,000, belonging to traders with fewer than ten establishments). (b) Sample coverage for small or new establishments of independent traders based on a sample of areas. In some areas there was then a full investigation, while in others a further sampling procedure was followed. In London, one shop in five in the central area was covered; of the seventy-six shopping centres in the rest of London documented in the 1961 Census, full coverage was given to a sample of thirty-three centres. In the remainder of London, six wards were chosen for full coverage, as extensive retail development had occurred. Remaining wards were sampled on a one in twenty basis, and were then fully covered. In sixty large towns, a sample of streets was covered (streets containing more than twenty small independent traders were divided into street segments, each containing between ten and twenty outlets). The sample was stratified by number of outlets, and by whether in a shopping centre or not (as defined for the 1961 Census) with varying sampling fractions. Thirty-two local authority areas with rapid development were covered by a sample of one shop in five. The remainder of Great Britain, consisting of about 1,680 local authority areas, was covered by a sample of local authority areas stratified by standard regions, urban/rural, conurbation/non-conurbation, and by total turnover in each local authority area. Sampling fractions varied, averaging one in ten. Within the sample areas there was full coverage. (For list of selected local authorities, see the Board of Trade Journal, 19/11/65, pages 1,232 – 1,233.) The sample totalled 85,000 shopkeepers, and gives regional figures of reasonable reliability. Unpublished statistics for the local authorities where there was a complete enumeration are stated to be available from the Board of Trade. First results published in the Board of Trade Journal, 23/2/68; remainder should be available in late 1969 (main volume) and early 1970 (floor space, transport costs, special forms of trading, sampling errors).

(iii) Annual enquiries in non-census years; these are based on small samples, and cover fewer topics.

3. Geographical Resolution

Basic information unit:

1950 – Questionnaires destroyed.

1957 – Sample of all largest retailers, and an average of 10% of small independent retailers (i.e. in all, about 12% of total retail). The questionnaires for establishments are subject to confidentiality restrictions (see 4. below), and the geographical resolution is restricted by level of sample reliability to national and some regional figures.

1961 – Questionnaires to retail organisations (establishments listed by them are coded by postal address) – these are subject to confidentiality restrictions: see 4. below.

1966 – Sample data – questionnaires to individual retail establishments (subject to confidentiality restrictions – see 4. below). Geographical resolution restricted to regions by level of sample reliability.

Processed data:

1961 – Retail organisations and establishments (coded by reference number). Subject to confidentiality restrictions – see 4. below.

1966 – Individual retail establishments – subject to confidentiality restrictions – see 4. below. Geographical resolution restricted to regions by level of sample reliability; however, some reliable data can be obtained for those local authority areas which were fully enumerated in the sample.

Published material:

1950 – Standard Regions, counties, Town Size (by population), Selected Towns (numbers of persons engaged and numbers of central offices and warehouses).

1957 – National figures only.

1961 – Conurbations, counties (summary totals only), local authorities, shopping centres.

1966 – National figures, Standard Regions.

4. Statistical and Legal Restrictions

Statistical limitations:

For the data which is based on samples (e.g. 1966), figures are not reliable below a regional level, except in the particular local authorities where a complete enumeration was undertaken. (See 2. and 3. above.) Some of the tables presented in published reports are adjusted to take account of non-response.

Confidentiality:

The Statistics of Trade Act prevents individual disclosures without permission of the business concerned. This has been reinforced; even the President's discretion has been limited by the Census of Distribution (Restriction on

Disclosure) Order 1960 (SI 1960 No. 2,364). Except for conducting the census or taking legal proceedings under the Act, individual returns cannot be revealed, even to other government departments. Small combinations of establishments are aggregated as long as there is no risk of revealing indications of individual returns. Where this is the case, the whole set of information is withheld, even including the number of establishments involved. With these provisos it is expected that data for small groups of establishments will be available by request when the 1971 results are available.

5. Boundaries

Local authority boundaries are available from HMSO or OS Standard Regions (HMSO) also. For boundaries of shopping centres, Board of Trade street lists must be examined.

6. Forms of Data Storage

Original data: In all cases, the original data consists of questionnaires, each with a postal address. For the 1961 Census, there are about 300,000 returns from organisations covering about 500,000 individual retail establishments. The original data for the 1950 Census have been destroyed: all that remains is the published material.

Processed data: Data is stored on magnetic tapes for use on a LEO 3 computer. In some cases summary print-outs have been prepared, one line per return. Each tape entry has a reference number (entered from the questionnaire), a local authority code, and a shopping centre or ward code in larger towns. Tapes are assembled randomly as returns are received.

7. Operational Details

Originating body: Board of Trade.

Location of supervisory staff: (a) Business Statistics Office,
Lime Grove, Eastcote,
Ruislip, Middlesex.

(b) Chartist Tower,
Dock Street,
Newport, Monmouthshire.
NPT 1 XG. (For general enquiries and 1971 Census matters.)

Frequency and dates of survey: Full census: 1950, 1961 (covered wholesale trade in 1950). Large sample census: 1957, 1966. Sample enquiries into stocks and capital expenditure are held in non-census years. Next full census: 1971.

Location of original data: See location of supervisory staff above.

Location of processed data: See location of supervisory staff above.

Publications: Published reports for 1950 and 1961 contain cross-analyses of establishments by numbers employed and size of turnover. For 1961 there are also summaries of the number of establishments selling particular commodities. Findings are summarised by old Standard Regions, incorporating estimates to cover non-response. Summaries are given for number of establishments, kind of business (nine major, twenty-two minor categories), turnover, and employment by conurbation, towns (over about 20,000), other areas, and major shopping centres. These centres were defined by the Board of Trade after consulting local authority officials. Definitions by street are available from the Board of Trade. For areas smaller than regions, figures are not adjusted for non-response. Any breakdown or analysis below a local authority and shopping centre level will certainly require a return to the original questionnaires. The 1971 Census is likely to be coded by 100-metre grid squares.

1. Subject **Industrial Development Certificates**

Content:

(a) Area of premises applied for (in square feet).

(b) Present and estimated additional employment. *

(c) Area of present unit, and its location. *

(d) Location of proposed premises.

(e) Type of industrial activity.

* The present employment and the area of the present unit are not included in the data which are processed.

Geographical coverage: Great Britain

2. Method of Data Collection

Under the Town and Country Planning Act, 1947, and subsequent consolidating Acts, planning applications for industrial premises must be accompanied by IDC's obtained from the Board of Trade, if the envisaged expansion is large enough to fall within certain prescribed limits. (These limits and their application to different parts of the country have changed considerably over time.) Following the initial application for an IDC made to the appropriate regional office of the Board, further details are usually obtained by letter, telephone or interview. Applications are made, approved or refused, and, if approved, contact is maintained until a project has been completed.

3. Geographical Resolution

Basic information unit: This varies between the accuracy of a map reference number and the less distinct local authority area. Normally, however, a postal address is given.

Processed data: Employment exchange area.

Published material: Economic Planning Regions.

4. Statistical and Legal Restrictions

Confidentiality: Because it is collected in confidence, the information about existing and estimated additional employment has particular protection from disclosure. Information about individual applications is regarded as confidential, and for publication purposes there are normally at least five cases per cell required as a minimum level of aggregation.

5. Boundaries

Standard Regions and Economic Planning Regions are available from HMSO.

6. Forms of Data Storage

Original data: The application forms carry the information listed in 1. above, together with a location reference for the proposed expansion. This reference may be a postal address and/or an ordnance survey map reference number, but in cases where actual sites have not been chosen, it may be just the local authority area which is stated.

Processed data: Monthly returns of applications (and the decisions on these which follow) are sent from the regional office to HQ Statistics Division for checking, before being punched onto cards at the Business Statistics Office (one application per card). Although Hollerith cards have only been in use for about a year, details of approvals from 1956 onwards, and of completions from 1964 onwards, have been repunched. These cards, of which there are about 70,000 (from 1956 to date), carry an employment exchange area code as the locational reference for the proposed development. To date, all tabulations have been produced by mechanical sorting of the cards; the use of magnetic tape is envisaged within the next year.

7. Operational Details

Originating body: Board of Trade

Location of supervisory staff: Statistics Division,
Board of Trade,
1 Victoria Street,
London, SW 1.

Frequency and dates of survey: These data are the result of a continuous process of control. However, published data relates to quarters of each year.

Location of original data: The application forms are generally retained at the regional offices of the Board.

Location of processed data: Business Statistics Office,
Lime Grove, Eastcote,
Ruislip, Middlesex.

Publications: Quarterly summaries of approvals by region are published in the Board of Trade Journal. Summaries of completions appear in the Monthly Digest of Statistics, and tables of applications and of estimated additional employment may be obtained by application to the Board of Trade (similarly, sub-divisional data is available to enquirers). Since August 1968, details of refusals (going back to 1960) have also been published.

Other Business Statistics Office Enquiries Undertaken by Staff at Eastcote

In addition to the main Census of Distribution and Census of Production, annual enquiries on those subjects are undertaken to fill in the information gaps between censuses.

Periodical sample enquiries are conducted to investigate a number of topics not covered by the two main censuses. These enquiries are primarily designed to provide information on stocks and capital expenditure. Large scale base year enquiries, held at five-yearly intervals, normally ask for turnover and its analysis, and are used to provide sampling frames for the enquiries in the intervening years.

The main subjects for these enquiries are:

Wholesale distribution
Motor trades
Property companies
Catering trade

Registered clubs
Taxis and private hire
Road haulage and road passenger transport
Construction
General (mainly entertainment, sport and recreation).

(Insurance and banking figures are provided by the Head Office of the Board of Trade.)

Business activities not covered by any of these enquiries include transportation (rail, air, water), GPO services, professional services, bookmaking and cleaning.

These annual enquiries offer only limited scope for planning/data bank purposes. They cover specialised topics and, because of the sample basis, are not usually suited to locational analysis, except at a very broad scale.

1. Subject — **General Traffic Census**

Contents: The large scale traffic censuses which this entry describes have been carried out every few years since 1922, most recently in 1965 (covering 2,200 points on the trunk road system), but with a new more continuous type of census having begun in 1969. (See 7. below.) The aim of these censuses is to gather information about variations in traffic flow from section to section of road for highway planning purposes. Traffic counts identify the numbers and types of vehicle passing each census point.

Geographical coverage: England and Wales and Scotland (in conjunction with the Welsh Office and the Scottish Development Department).

2. Method of Data Collection

Counts are made by observers at each census point. Separate directional counts are taken where the traffic volume necessitates them, but the results are combined on the summary tapes. Tally books are used to record hourly numbers of vehicles in each of five categories (two-wheeled motor vehicles, cars/taxis, buses/coaches, light vans, other goods vehicles) and bicycles are also counted where they are important. (Earlier censuses used eleven categories, two-wheeled motor vehicles and other goods vehicles being subdivided.) The survey points are not random; they are selected to provide information on each main link in the trunk and principal road network (as far as possible). The same sites are used in each survey, unless new road construction makes a site redundant. Completed tally books go to Hemel Hempstead for preliminary scrutiny; the books are read optically on to tape at HMSO Norwich; from this a summary tape, giving sixteen-hour daily totals for each class of vehicle, is prepared; estimated traffic on an average August day is calculated for each point, and results printed out together with rates of growth (compared with 1965 – trunk road points only).

3. Geographical Resolution

Basic information unit: Tally books for each census point. (However, these are not used as basic data files – see 6. below.)

Processed data: Summaries for census points.

Published material: Data for about one-third census points (1965 Survey).

4. Statistical and Legal Restrictions

Statistical limitations: The Census is designed to measure the flows on each link of the trunk and principal road network. Data for individual sites should be treated with care, because of the limited period of counting, and they do not necessarily represent conditions at nearby sites.

Confidentiality: No restrictions.

5. Boundaries

This is not applicable, since point data and line flows are involved. The MOT has a series of maps with the census points marked on them: they are available from St. Christopher House.

6. Forms of Data Storage

Original data: Tally books, each identified by a census point number. However, these are not used as a practicable data file, and the data are stored on summary sheets or magnetic tape in point number order. For the cycle of counts starting in 1969, a taped list of points with eight-digit grid reference numbers will shortly be available.

Processed data: For the three surveys up to 1965, a summary tabulation sheet was produced for each tally book, incorporating the census point number, and showing traffic volumes by vehicle category by hour of day. The 1965 survey resulted in about 22,000 summary tabulation sheets, one per site per day, and about 6,000 where directional flows were not combined. Data from 1969 onwards are transferred to magnetic tapes, and these contain the smallest unit of information available (i.e. vehicle class/hour).

7. Operational Details

Originating body: Ministry of Transport

Location of supervisory staff: St. Christopher House,
Southwark Street,
London, SE 1.

Frequency and dates of survey: Post-war general traffic censuses have been conducted as follows:
1954 – trunk roads (1,500 points) and Class 1 roads (4,500 points)
1961 – trunk roads (1,750 points)
1965 – trunk roads (2,200 points)

In 1954 and 1961 counts covered 7 x 24 hours in August. In 1965 counts covered 4 x 16 hours in August at about two-thirds of the sites; at the remaining sites counts were taken for 4 x 24 hours in August and 3 x 24 hours in each of the months of February, May and November.

Starting in August 1969, there is an annual 2,000-point census to form part of a four-year cycle totalling 8,000 points on motorways, trunk and principal roads. Counting covers 2 x 16 hours (Sunday and Monday) in August, with additional two-day counts at five hundred of the sites in April/May.

Location of original data: MOT Data Processing Branch,
Hemel Hempstead, Hertfordshire.

Location of processed data: A full set of summary results are held by the Road Traffic Census Unit, Directorate of Statistics, St. Christopher House. Magnetic tape for 1965 data and growth rates are held by the Ministry of Transport Mathematical Advisory Unit, Victoria Street, London, SW 1. Magnetic tapes for 1969 onwards will be available on application to the Census Unit.

Publications: Summary results are presented in map form in "Highway Statistics", 1967, based on data from about one-third of the census points used in 1965. Data from the 1969 survey are still being analysed, but are likely to be published in a future edition of "Highway Statistics".

1. Subject **Continual Sample Surveys of Road Traffic: (all roads except un-classified)**

Content: This provides estimates of traffic flows (by eleven classes) to be grossed up to represent the total vehicle mileage throughout the country, month by month. It is used mainly to assess seasonal patterns and long-term trends (see General Traffic Census for estimates of flows along particular roads).

Geographical coverage: England and Wales and Scotland (in conjunction with the Welsh Office and the Scottish Development Department).

2. Method of Data Collection

These surveys have been operated since 1956, starting with automatic counters at fifty sites on the trunk and classified roads of Great Britain; the sites were chosen on a statistical basis so as to make them representative of all classified roads in both urban and rural areas. Starting in April 1957, a series of manual counts was introduced, and these are made at the same sites. The counts are spread over the different weekends in the month, excluding bank holiday weekends when special counts are made at some of the sites. Counts are made by the local authorities who are responsible for the maintenance and reading of the automatic counters. In 1966 the number of sites for manual counts was increased to 200. The 200 points are a random sample, unchanged from year to year, stratified by four classes of road (trunk, Classes 1, 2 and 3) and by type of local authority (urban, rural). The thousand-mile motorway network was included in the original sample, and is included as sections are built. Flows in each direction are recorded separately where the traffic volumes necessitate this.

The tally books of counts (one book per site per day, or two if the directional flows are retained) are passed to Hemel Hempstead. A logabax machine (a sophisticated adding/storage machine) is used to produce summary tabulations of numbers by each class of vehicle by hour of day. The total flow in each category is multiplied by factors so as to produce daily, weekly and monthly estimates of vehicle miles run on the whole trunk and classified road system.

3. Geographical Resolution

Basic information unit: Tally books for each census point (geographical resolution restricted by level of sample reliability).

Processed data: Summaries for census points (geographical resolution restricted by level of sample reliability). Grossed-up national summaries for each type of road.

Published material: National figures only.

4. Statistical and Legal Restrictions

Statistical limitations: The automatic counters at fifty sites yield, in principle, a continuous record of traffic flows (apart from counter breakdowns). With the manual counts, the restriction of counting to a few days a month makes the results for individual points liable to errors due to:

(a) Chance variation in traffic flows from hour to hour, week to week, etc.;

(b) The use of scaling factors from short counts to annual totals: these may not be appropriate to particular points or to an average point.

The national estimates of trend in vehicle mileage, which are the end purpose of the counts, are subject additionally to sampling errors due to the relatively small number of points.

Confidentiality: No restrictions.

5. Boundaries

This is not applicable since point data and line flows are involved; census sites are marked on 1" maps held at St. Christopher House.

6. Forms of Data Storage

Original data: Tally books or logabax sheets giving equivalent detail.

Processed data: Summary tabulations produced by logabax machine.

7. Operational Details

Originating body: Ministry of Transport

Location of supervisory staff: St. Christopher House,
Southwark Street,
London, SE 1.

Frequency and dates of survey: 1956 onwards – counters recording continuously at fifty sites.
1957 onwards – manual counts at weekends at the same sites.
1966 onwards – manual counts at weekends at two hundred sites.

Location of original data: MOT, Data Processing Branch,
Hemel Hempstead, Hertfordshire.

Location of processed data: Master copies at Hemel Hempstead, but also available at St. Christopher House.

Publications: Monthly figures are published in index form in the Monthly Digest of Statistics; vehicle mileage figures are given in Highway Statistics (HMSO).

1. Subject **Surveys of Road Goods Transport**

Content:

1952

(i) Characteristics of goods vehicles: Information collected on unladen weight, carrying capacity, type of carrier's licence and type of fuel.

(ii) Operations of goods vehicles: Mileage run loaded to more than half capacity, less than half, or empty, and maximum load for each journey in survey period.

1958

(i) Characteristics of goods vehicles: Information collected on unladen weight, carrying capacity, type of carrier's licence and type of fuel. (Nature of business of C-licence operators also known.)

(ii) Operations of goods vehicles: Whether journey straight-run or stopping, mileage run, loaded to more than half capacity, less than half, or empty; maximum load on each journey; goods carried on each journey in survey period.

1962

(i) Characteristics of goods vehicles: Information collected on size of fleets, numbers of articulated vehicles and containers in fleets, licence type and type of fuel of each vehicle, body type, articulation and carrying capacity. For C-licence vehicles, nature of business of operator and type of work vehicle engaged on in survey period. For B-licence vehicles, whether used mainly for hire and reward or on own account in survey period. (Garage address also known.)

(ii) Operations of goods vehicles: Day of week, origin, destination, mileage, tonnage, whether straight-run or stopping, carrying capacity of any trailer used, and goods carried for each journey (except C-licence vehicles under one ton).

1967 – 1968

(i) Characteristics of goods vehicles: Information collected on unladen weight, carrying and type of fuel, body-type, articulation, whether usually carries demountable container, number of axles, year of first registration. For C-licence vehicles, nature of business of operator and type of work vehicle engaged in during survey period. For B-licence vehicles, whether used for hire and reward or on own account during survey period. (Licence-type and garage address also known.)

(ii) Operations of goods vehicles: Day of week, origin, destination, mileage, tonnage, whether straight-run or stopping, whether extended overnight, use of shunt drivers, changeover of drivers, use of containers for each journey.

Coverage:

(i) Type of vehicle: 1952, 1958, 1962, 1967/8; General goods vehicles subject to carrier's licencing. (Excludes certain categories in 1952.) In 1958, 1962, 1967/8 Crown vehicles and agriculture vehicles excluded.

(ii) Geographical: 1952, 1958, 1962 and 1967/8; Great Britain.

2. Method of Data Collection

1952, 1958, 1962 and 1967/8 by mail questionnaire sent to operators of a sample of vehicles (using statutory powers of Statistics of Trade Act, 1947).

Sample unit: Vehicle (in one week)

Sample size: 1952 – 7,000 vehicles
1958 – 7,000 vehicles
1962 – 40,000 vehicles, divided between four quarterly surveys.
1967/8 – 49,000 vehicles, divided between twelve monthly surveys.

Sampling frame: 1952, 1958, 1962, 1967/8 Goods Vehicle Index. The Index is a list containing all vehicles on carrier's licence, originally kept on binders in thirteen Traffic Area Offices. Transfer onto punched cards, begun in 1961 for the 1962 survey, was completed by 1963. File then contained records for all A, Contract A, B-licence vehicles and C-licence vehicles over three tons (400,000 records). Provision was made for monthly up-dating. In 1967, the file contained 500,000 records.

Stratification: 1952, 1958, 1962, 1967/8 by licence-class, area and unladen weight. In 1967/8 variable sampling fractions were chosen to optimise accuracy. A post-selection subdivision of unladen-weight strata was used in analysis.

3. Geographical Resolution

Basic information unit: 1958, 1962 and 1967/8 – Questionnaires contain address of respondent. 1967/8 questionnaires contain garage address of vehicle, origin and destination of each journey. These cannot be made available to outside users because of confidentiality conditions of Statistics of Trade Act.

Processed data: 1952, 1958 – None
1962, 1967/8 – Garage address, origin and destination of each journey coded to zones. These cannot be made available to outside users because of confidentiality conditions of Statistics of Trade Act.

Published material: 1952, 1958 – No geographical breakdown.
1962 – Origins and destinations of journeys analysed by old Standard Regions. Classification by smaller groups of local authorities used to map freight movements.
1967/8 – Location of vehicles, origins and destinations of journeys analysed by new Standard Regions. Number of vehicles analysed by MOT "Standard Classification for Traffic and Transport Data" at three-digit level.

4. Statistical and Legal Restrictions

Statistical limitations: 1962 – Relatively large sampling errors are present in the geographical analyses. No information was collected on origin, destination of journeys made by C-licence vehicles not over one ton. These were assumed to be within zone of garage address.
1967/8 – Small sample numbers will produce relatively large sampling errors in geographical analysis. Ton-mileage figure for stopping journeys, value of goods carried, are estimates only.

Confidentiality: Under the Statistics of Trade Act, 1947, individual returns cannot be made available without permission of respondent. Aggregate data for MOT zones from 1967/8 survey can be made available for all classes of sufficient size so that individual operators cannot be identified.

5. Boundaries

1962 – Local authority-based zones, subdividing Standard Regions. Definition in MOT "Standard Classification for Traffic and Transport Data".

6. Forms of Data Storage

Original data: 1952 – Destroyed.
1958 – Questionnaires exist, but see 4. above.
1962 – One-tenth of questionnaires exist, but see 4. above.
1967/8 – Questionnaires exist, but are to be destroyed.

Processed data: 1952, 1958 – None.
1962 – Card records as follows: (a) Vehicle and journey cards, (b) Commodity analysis data, (c) Geographical analysis data, and (d) Origin and destination matrices.
1967/8 – Card records containing coded data from questionnaires and edited magnetic tape including grossing factors. The tape contains data coded geographically to three-digit level of MOT "Standard Classification for Traffic and Transport Data".

7. Operational Details

Originating body: Ministry of Transport

Location of supervisory staff: Statistics Directorate of Ministry of Transport,
St. Christopher House,
Southwark Street,
London, SE 1.

Frequency and dates of survey: 1952 – Week ending 28/9/52
1958 – Week ending 27/4/58
1962 – April 1962 to January 1963
1967/8 – June 1967 to May 1968

Location of original data: 1952 – Destroyed
1958 – St. Christopher House
1962 – St. Christopher House
1967/8 – Ask St. Christopher House (see 6. above).

Location of processed data: 1952 – None
1958 – None
1962 – MOT ADP, Harmondsworth
1967/8 – Ask St. Christopher House.

Publications: 1952 – JRSS 117 pages 297-330 (1954)
1958 – MOT & Civil Aviation : "The Transport of Good by Road" (1959), and see JRSS (A) 123 (1960)
1962 – MOT Final Results Part 1 (1964)
Commodity Analysis (1965)
Geographical Analysis (1966)
Methodological Report (1966)
1967/8 – To be published by MOT

1. Subject **National Travel Survey**

Content: The Survey covers characteristics of (a) households, (b) individuals in households, (c) vehicles available, (d) journeys made, (e) stages of journeys.

Geographical coverage: Great Britain.

Method of data collection: The sample consists of persons of three years of age and over, clustered by household. The sample households are chosen by a two-stage operation using the register of electors of Great Britain as a frame. In the 1964 survey the primary units in England and Wales were wards in the urban strata, and groups of contiguous parishes in the more sparsely populated rural areas. In Scotland the units were wards, small burghs or electoral divisions. In England and Wales the primary units were stratified by type of area, standard region and the 1955 Juror Index developed by the Government Social Survey as a means of socio-economic classification of areas. In Scotland the units were initially stratified into four main geographical areas, and then by rateable value per elector. The second stage was the selection of addresses with equal probability from each primary unit. In 1965 the primary units were local authority administrative areas, followed by the two stages used in 1964.

Stratification was by planning regions, standard regions and type of area in England and Wales, with an extra strata of geographical sector introduced for Scotland. The 1966 survey was sampled in a similar manner to 1965.

A travel record diary was distributed prior to the commencement of the recording period to all informants. Each informant was interviewed at the end of his/her recording period, and a detailed schedule of all journeys was completed by the interviewer, the informant using the Travel Record as an aide memoire. The household data were collected at the initial interview with the head or other responsible member of the household.

The 1964 survey yielded 5,000 respondent households. With this sample size, areal analysis below national level is impossible. The 1965 survey had a much larger sample basis and covered 12,500 households. One of the aims (in 1965) was to have about 1,000 households in each region, but because East Anglia had at the time only just become an economic planning region, separate consideration was given to the area, in that it was stratified, but it was nevertheless under-sampled. In fact, the sample only covered about 350 East Anglian households (and therefore about 1,000 individuals and about 11,250 journeys) and regional analysis of the data is considerably less reliable than for other regions.

The 1966 survey had a reduced sample covering about 5,250 households. It is envisaged that the next National Travel Survey (1971) will return to the large sample size of about 15,000 households – producing usable results from about 12,000.

3. Geographical Resolution

Basic information unit: Destroyed

Processed data: Data for individuals in sampled households including details of trips made (geographical resolution restricted by level of sample reliability). Summaries for regions (1965 survey only, and excluding East Anglia).

Published material: Mainly national figures: some regional summaries.

4. Statistical and Legal Restrictions

Statistical limitations: Of the surveys undertaken so far, only that for 1965 had a large enough sample to produce reliable results at the regional level. A statistical report on the 1965 survey is being prepared by MOT for publication (in 1971); this will, amongst other things, present a general indication of the size of statistical errors involved in the several hundred tabulations produced.

Confidentiality: Confidentiality is not a problem; individual data units are not identifiable.

5. Boundaries

Economic Planning Regions.

6. Forms of Data Storage

Original data: Original data are in the form of questionnaires, which are filled in by the interviewers employed on behalf of the Ministry of Transport. For the 1965 survey, data were obtained for 12,500 households, 34,000 individuals, 7,000 vehicles and 400,000 journeys. These original documents have all been destroyed.

Processed data: Data are transferred from original questionnaires to punched cards, and the processing is carried out by a private computer bureau. For 1964 and 1965, the work was done on a CDC machine, but for the 1966 data, processing was carried out on a new machine – a UNIVAC 1108. 1965 survey information is now held on four edited data tapes which replace the original fifteen raw data tapes. In addition, the raw data have been sorted into three parameter tapes as follows:
Tape 1 – Household and Vehicle Characteristics
Tape 2 – Household and Individual Characteristics
Tape 3 – A complete data file (much larger than tapes 1 and 2).

7. Operational Details

Originating body: Ministry of Transport

Location of supervisory staff: Statistics A Division,
St. Christopher House,
Southwark Street,
London, SE 1.

(See also Location of processed data.)

Frequency and dates of survey: 1964, 1965, 1966: Future decisions about whether to carry out the survey again are dependent on (amongst other things) the available resources of the Ministry of Transport and the completion of analysis of existing results. However, it is probable that the next survey will be in the year 1971.

Location of original data: Destroyed.

Location of processed data: The 1964 magnetic tape is held by the Ministry of Transport's ADP Section. The 1965 and 1966 magnetic tapes are still held at the computer bureau – Scientific Control Systems Ltd.

Publications:

Results of the National Travel Survey have not normally been published in full. The 1967 volume of Highway Statistics has a number of tables describing characteristics of household vehicles, from the 1965 survey (pages 35 – 38). The 1967 volume of Passenger Transport in Great Britain also has a series of tables describing characteristics of passenger journeys.
A five-volume technical report covering the three years has been produced for internal use within the Ministry. A statistical report on the 1965 survey will be published in 1971.

1. Subject **Road Accident Statistics**

Content:

A pre-coded form covering casualties, vehicles involved, drivers and riders, attendant circumstances (e.g. weather, road and lighting conditions, road class, speed limit, etc.) is completed as detailed in 7. below. In most cases this is punched onto 63 columns of an 80-column card. A space of 6 columns is left for a locational figure code and, if used in conjunction with the 3-column local authority code, gives accuracy to the nearest 100-metre. Some large local authorities have elected to use 8 columns (i.e. extending to column 65) to give greater accuracy (e.g. London). A third possibility for locational information is used in four counties in the North West (Cheshire, Lancashire, Westmorland, Cumberland). This is a Road Mileage Reference Scheme working on the principle that road accidents, by definition, happen on roads, and therefore a higher degree of accuracy can be more economically achieved by only taking these into account.

The Police Force Area is also coded on the card but, with the recent amalgamation of forces throughout the country, these are often larger in size than counties.

Geographical coverage: Great Britain (see 2. below).

2. Method of Data Collection

The information on the pre-coded forms is punched onto cards and then transferred to magnetic tapes and processed on the Road Research Laboratory computer at Crowthorne. The present form (Stats. 19) and its accompanying booklet (Stats. 20) were introduced on 1st January 1969. Before then, the only locational information was a local authority code (RD, UD/MB etc.). Magnetic tapes are available for data back to 1953. The accident forms are sent in by the police a month at a time.

Scotland punches all its own information onto cards, and the Scottish Development Department in Edinburgh forward the information to the Ministry of

Transport. In England and Wales, the individual police forces send their data to the Ministry of Transport at Hemel Hempstead.

3. Geographical Resolution

Basic information unit: Data for individual accidents coded to grid reference or mileage reference on particular roads.

Processed data: Details of individual accidents on tape. Police force summaries.

Published material: Monthly figures on a national basis.

4. Statistical and Legal Restrictions

Statistical limitations: The only limitation is that the survey's coverage is limited to those accidents which are reported to the police, and which involve personal injury or death.

Confidentiality: There are no confidentiality problems, since names or addresses of persons involved in the accidents are not recorded.

5. Boundaries

Local authority and police force area boundaries are available at MHLG.

6. Forms of Data Storage

Original data: 80-column cards contain nearly all the information on the accident form. The day of the week is not included, and must be derived from the date. If persons under fifteen are involved, a brief account is written on the form or on an attached sheet; this cannot be coded or used statistically, but is of value to RoSPA in its work with children.

Processed data: Information is aggregated to police forces, which, since amalgamation, often cover more than one county. Tabulation of data by CB, MB, UD, RD, Scottish Burgh, etc. is feasible, although special programming is required.

7. Operational Details

Originating body: Ministry of Transport (in conjunction with Scottish Development Department).

Location of supervisory staff: MOT,
St. Christopher House,
Southwark Street, London SE 1.

Frequency and dates of survey: A form (Stats. 19) is filled in by the police for each reported road accident on the public highway resulting in personal injury or death.

Location of original data: About half the original forms are kept by the police forces who punch the information onto cards and send them to the Ministry. Other police forces send their forms to the Ministry processing branch at Hemel Hempstead which punches the cards.

Location of processed data: Cards at Hemel Hempstead are kept for about ten years. The magnetic tapes are at the Road Research Laboratory at Crowthorne, Berkshire.

Publications: Monthly press notice. Annual Reports (entitled "Road Accidents") – no area classifications. RoSPA publish some figures by police force area. (See last sentence of 1. above.)

1. Subject **Statistics of Licenced Motor Vehicles**

Content: The licence renewal form seeks information on registration mark, taxation class, make of vehicle, name and address of vehicle owner, place where vehicle is normally kept, present registration authority, and two additional questions about vans and dual-purpose vehicles.

Geographical coverage: Great Britain

2. Method of Data Collection

Detailed figures for counties and county boroughs are aggregated at present by means of a logabax machine (a sophisticated adding machine with the ability for storing numerous items of information). Plans are in hand for a completely new computer installation in Swansea, and provision is made in the new Bill (see 7. below) for all registration to be centralised. The take-over will begin in 1974, and it is hoped that it will be completed by 1977. The computer will begin by dealing with new registrations, and then will gradually take over current licences. It is possible that a decision will be taken to record the place that the vehicle is kept, as well as the address of the owner (both questions are asked on the present form). When the computer is functioning, rural district and urban district data will be available (in London, postal districts). Location coding by grid squares is considered by the Ministry of Transport to be impracticable with the large number of vehicles involved.

3. Geographical Resolution

Basic information unit: Licence file for each vehicle registered.

Processed data: Summaries for counties and county boroughs

Published material: Summaries for counties and county boroughs

4. Statistical and Legal Restrictions

Confidentiality: In principle there are no restrictions; pending full computerisation of records, only limited analyses are available.

5. Boundaries

Present material relates to counties and county boroughs: boundaries are available on a standard map.

6. Forms of Data Storage

Original data: Forms are completed when a vehicle is licensed, and at present contain both the owner's address and the place where the vehicle is kept, if different. It is not certain whether place of keeping will be retained after licensing become centralised.

Processed data: Tabulations arrived at by logabax machine are annual summary sheets for major local authorities.

7. Operational Details

Originating body: Ministry of Transport

Location of supervisory staff: MOT,
St. Christopher House,
Southwark Street,
London SE 1.

Frequency and dates of survey: An annual census is taken on a 10% sample basis of all vehicles that had current licences in the third quarter of the year. After 1976, when all licensing will be centralised at Swansea, data for all vehicles will be held on a computer file and censuses will be possible several times per year.

Location of original data: Records are retained in the local authority Tax Office which issues the licence. If the car is later registered in another area, the file of past licences is sent from the former local Taxation Office to the new one.

Location of processed data: There is no computer processing at present. The logabax machine work is carried out at Hemel Hempstead. The centralised computer will be located at Swansea (see 2. above).

Publications: Highway Statistics (annual) gives census figures for economic planning regions, and counties and county boroughs within these. New registration statistics for counties and county boroughs are published quarterly in the Business Monitor Series.

1. Subject **Incomes Survey**

Content: Basic records contain characteristics of employment, pay, and details of allowances, dependants, etc. They also include addresses of residence and workplace administration. However, the income surveys have not hitherto taken account of the home addresses of individuals. Plans to do this in the future are under consideration. (See 8. below.)

Geographical coverage: United Kingdom

2. Method of Data Collection

Hitherto the area analysis has been made by reference to the General Claims District (GCD) of the individual, which is determined by his main source of income, e.g. a wage or salary is dealt with by the tax district dealing with his employer's PAYE scheme. The area analysis may, therefore, be inaccurate for those taxpayers who live or work outside their GCD. The analyses by standard region and county are built up by combining tax districts, some of which may have to be split for this purpose.

Within each district, a systematic sample of cases is taken within each category of taxpayer, e.g. Schedule D and various types of Schedule E (PAYE) taxpayers. Sampling is more intense in the higher income ranges. Forms completed in respect of each taxpayer sampled are processed on the Board's LEO 3 computer at Worthing. The resulting aggregates are corrected for known deficiencies (e.g. large formal arrangements for the decentralisation of tax records) and scaled to fit separately available national total figures.

3. Geographical Resolution

Basic information unit: Tax records for individual persons, including address of workplace (subject to confidentiality restrictions – see 4. below).

Processed data: County summaries. Geographical resolution is restricted by level of sample reliability.

Published material: Annual surveys produce national and regional figures.
Quinquennial surveys produce national, regional and county figures.

4. Statistical and Legal Restrictions

Statistical limitations: The sampling frame provided by tax records excludes many incomes below the effective tax exemption limit. In the regional and county analyses of survey data, the Armed Forces and Merchant Navy are treated separately. Estimates are subject to sampling error.

Confidentiality: Information provided for tax purposes is regarded as highly confidential. Great importance is attached to the confidence that has been built up with the public, and it is thought undesirable that basic data collected for tax purposes should be used for any secondary purposes in an unaggregated form. In preparing income distributions, any cell containing less than ten cases is either suppressed or amalgamated; this is most common in the upper income ranges.

5. Boundaries

There is no formal map showing tax district office boundaries. County and Economic Planning Region boundaries are available from HMSO.

6. Forms of Data Storage

Original data: The incomes classified are in terms of income tax units, in which a married couple, including dependent children, is counted as one. The files at local offices cover about twenty-five million persons. The return for the annual sample contains a 3-digit area code referring to counties and region, but no home address.

Processed data: An increasing proportion of tax records is held at the computerised regional centres. Sample survey records on magnetic tape for LEO 3 computer include the 3-digit area code, and from 1967/8 onwards, the tax district code.

7. Operational Details

Originating body: Board of Inland Revenue

Location of supervisory staff: Somerset House,
London WC 2.

Frequency and dates of survey: The survey of personal incomes takes place quinquenially (sample size – one million cases; first survey 1949/50) and annually (sample size – 125,000 cases; first survey 1962/3).

Location of original data: Held at 712 tax district offices.

Location of processed data: The PAYE tax records which relate to the incomes and personal circumstances of people with wages or salaries will be transferred to computers at nine regional centres over the next ten years. The first centre at East Kilbride, covering Scotland, will be fully operational by April 1970. There are also plans to transfer Schedule D assessing (of the trading profits of self-employment) to these computers; and plans are being made to transfer surtax assessing to a computer at Worthing in 1971. Processing of the sample records involved in the survey is handled at Worthing.

Publications: Annual reports of the Commissioners of Inland Revenue.

A Short Note on Company Taxation

Companies are liable to corporation tax on their whole income, and also on their realised capital gains. Their payments of dividends and interest are made under deduction of income tax which they have to account for to the Inland Revenue under Schedule 12. The unit for a corporation tax assessment is the company which may involve numerous establishments and a wide variety of activities. A company's tax affairs are dealt with in the district of its head office, and that, of course, may not coincide with the location of its establishments, which may be spread to a number of other tax districts.

Corporation tax assessments are analysed (at Worthing) according to the Inland Revenue trade classification, which differs in minor respects from the SIC. Multi-activity companies are allocated on the basis of the main activity, or what is normally their main source of profit (in contrast to allocation by main output by the Board of Trade, and main employment by the Department of Employment and Productivity). Corporation tax assessments are not analysed regionally. When the Central Register of Businesses is set up, a hierarchical indexing system will be devised to classify companies, enterprises, establishments, departments, etc., all within the SIC, and by area. Even when this is established, there will remain the intractable problem of allocating the single figure of profits of the heterogeneous company between its different areas of activity and trade groups.

1. Subject — **Rateable Value and Floor Space Assessments**

Content: Rating records contain descriptions of hereditament (e.g. first-floor flat, plus address) use, and assessed value. Nearly all private and local authority buildings, gardens, parks and playing fields are assessed for rating purposes, although certain categories, e.g. buildings for religious purposes, are exempted. Farmland is not assessed, and full rates are not payable on vacant property. The floor space survey is concerned with floor space assessed for rating purposes in selected use classes (i.e. commercial offices, Central Government Offices, Local Government Offices, shops and restaurants, factories, workshops and warehouses).

Geographical coverage: England and Wales

2. Method of Data Collection

Valuation officers and their staff visit or obtain plans of building work notified to them by rating authorities. Assessments are made of changes in value due to alterations, largely based on changes in floor space. Assessed values are recorded in valuation lists arranged generally in street order. These are kept up to date by a set of records of reassessments forming an annex. Complete sets of these records are on display at the offices of local authorities.

3. Geographical Resolution

Basic information unit: Assessments for individual hereditaments based on plans, visits, etc.

Processed data: Rateable Value – Valuation list gives assessments for individual hereditaments. Floor Space – Estimates of floor space for local authorities are prepared for MHLG.

Published material: Rateable Value – Local authority totals are in publications of IMTA (Institute of Municipal Treasurers and Accountants).
Floor Space – By local planning authorities, sub-divisions and regions.

4. Statistical and Legal Restrictions

Confidentiality: Rating assessments are available to the public, and the contents of the valuation list are not confidential.

5. Boundaries

Plans and survey details relating to individual hereditaments are held by valuation officers, and these are confidential documents. The arrangement of the records is by local authority, subdivided by rating districts, wards, etc.

6. Forms of Data Storage

Original data: Plans and survey details.

Processed data: Registers of assessments held at local authority and valuation offices, arranged by street. Summaries of floor space estimates by major land use by local authorities are prepared by the local valuation offices, and passed to the Ministry of Housing and Local Government, via the Inland Revenue Central Office. Any analysis below local authority level would require the use of registers of individual hereditaments.

7. Operational Details

Originating body: Commissioners of Inland Revenue.

Location of supervisory staff: Finsbury Square House,
London, EC 2.

Frequency and dates of survey: Rateable value of individual properties is re-assessed when there is a change of circumstance (e.g. a conversion or extension) or when a disagreement about the existing valuation results in an amendment. Revaluation of all properties takes place periodically to take account of the trends in rental values; the last revaluation came into force on 1st April 1963, and the next is due to take effect on 1st April 1973. Since 1st April 1964, estimates of floor space have been made from the rating records in selected use classes, and each year changes in floor space have been recorded in these classes. On 1st April 1967, an estimate of total floor space in selected use classes was made.

Location of original data: At local valuation offices.

Location of processed data: Local authority floor space summary data are passed to the Ministry of Housing and Local Government. Valuation lists and records are held in local authority offices, and in the local valuation offices.

Publications: (i) General estimates of floor space changes have been included in the Annual Report of the Commissioners of Inland Revenue. From 1969 onwards, details of floor space changes (and a report of the 1967 total count) will be published by HMSO, entitled "Statistics for Town and Country Planning", Series 2, No. 1.

(ii) "Rates and Rateable Values – England and Wales": published annually by HMSO.

Post Office

Telephone installations: The GPO publishes an annual booklet ("List of Exchanges – Part 1") which gives details for each exchange, of the number of "connexions", broken down by business and residence, and also gives numbers of actual "stations" for each exchange. In the language of the GPO, a "connexion" means a telephone line from a subscriber to the exchange (rather than a connected telephone call)and a "station" means an actual telephone. Thus, a private house with one telephone number and one telephone line to the exchange, having three extensions within the house, would rank as one "connexion" and four "stations".

Telephone calls: With the introduction of the Subscriber Trunk Dialling system (STD), all telephone calls will be recorded and charged in terms of 2d. units, the length of time bought for 2d. varying according to the distance of the call and the time of day. Each subscriber has a meter at the exchange which registers the number of 2d. units used, and this makes it impossible to identify direction, frequency or duration of calls. This has two consequences:

(i) Telephone bills do not give details of individual calls (except for those made through the operator);

(ii) It is not possible to perform statistical analysis of calls in terms of any variable other than 2d. units, and this prohibits all locational analysis.

It is, however, necessary for the GPO to have information about calls from time to time, and they concern themselves particularly with the "busy hour", i.e. the peak period for telephone calls. On main telephone routes from one exchange to another, they install overflow meters which register each time calls cannot get through because all the lines are engaged. By this means, they can estimate to what extent line occupancy is overloaded, and where improvements would be appropriate.

There are also facilities by which the GPO can assess usage of lines. This is done by a sampling method, and enables them to determine line occupancy and routing of calls.

Postal services: A separate entry in the catalogue covers the postal coding system being introduced by the GPO (see next entry in this Appendix). Post Offices are of two types, crown offices and agency offices, and whilst there are 2,000 crown offices in the country, the 23,000 agency offices conduct 50% of the postal business of the GPO.

Crown offices keep records of transactions for the purpose of weekly balancing. Accounting is done in four or five week accounting periods, ending on the first Friday in the month. This only takes note of the value of stamps sold, not the numbers by denomination. An accounting office in Edinburgh deals separately with Scotland.

There is a classified count of mail anything up to once every five years (e.g. Brighton do a classified count of mail every year).

Parcels: One bag in five hundred is examined in detail, and figures are sent monthly to headquarters. GPO regions are then given a regional parcel figure. In addition to this sample, there is an annual count one week in October of parcels sent through Crown post offices (this may be suspended by regional offices in four out of every five years).

Letters: Mail bags of letters pass directly between offices if there is sufficient demand, or via collecting and distributing offices if numbers are small. It is extremely difficult to estimate exactly how much mail is in any particular bag, and how many bags pass from one postal area to another. Franking machines at local offices can provide counts of some outgoing letters, but with no directional detail.

Postal Codes

The GPO is gradually introducing a system of coding groups of addresses to facilitate automatic sorting of mail. Norwich was the first city in which this scheme was implemented, but the early types of code used were rather different from the present codes.

The Norwich experiment started in December 1958, and ended in 1960. Based on the success of the Norwich experiment, the Post Office decided to extend postal coding to the rest of the country, and in 1966 the Croydon postal coding was launched, followed by Aberdeen in early 1967. By the early 1970's the seventy-five largest post towns in the country will have their postal codes, and this will represent well over 75% of all the mail handled by the British Post Office.

The precise significance of each of the six characters of the code require defining, because there are some changes in significance arising from the slight changes made after the initial coding installation was made at Norwich. According to Post Office officials at their South Eastern Region headquarters in Brighton, the usual pattern is to have six alpha-numeric characters in two sets of three, the third and fourth characters being numbers and the remainder letters. A typical code is shown below:

BN1 2EQ

Meaning of the symbols: BN – Brighton

1 – A postal area served by Brighton (bearing no relation to other administrative boundaries).

2EQ – This represents the destination for delivery and may be:

(i) One establishment or firm (only applied if they have twenty or more letters delivered per day on average);

(ii) One address – this may be the same as (i) above, or it may be one address with several firms;

(iii) A section of a street (with roughly twenty-five possible calls for the postman); when the codes are set up, fewer than twenty-five calls will probably be included to allow for expansion of residential or commercial activity;

(iv) A whole street, providing there are not too many calls within it;

(v) A number of small streets, e.g. a cul-de-sac, such as Williton Close, may be included with the road onto which it gives access, say Williton Avenue;

(vi) The second group of characters can be repeated in other areas bearing a different number in the first group of characters, e.g. BN2 2EQ. Ultimately there can be ten sectors in each area represented by the number in the first group of characters.

Thus, the postal codes do not correspond directly to postmen's walks, which are typically about four hundred houses in a residential area. Whilst such a direct correlation might have seemed logical, the system used is eminently more flexible, since postmen's routes can be aggregated in a variety of ways and can easily be changed if the need arises.

The suggestion has been made that the postal coding system might, with the addition of a house or flat number, prove to be a unique locational reference and thereby offer itself as an alternative to co-ordinate references, grid squares or street blocks. In fact, because of (v) above, this is not so, as, say, 3 Williton Close and 3 Williton Avenue would have exactly the same code.

It is hoped that the scheme will be complete by the early or mid-1970's, but even then it will not be a "carpet" cover for the whole of the country, as many rural areas will not be included. The object of the codes is to ensure that letters finish up with the right delivering postman. Thus, the codes refer to sections of public highways open to postmen, and are used to identify the letter box or boxes in the vicinity. It would involve a significant extension of the system to cover all land, especially in rural areas, as it would not be clear which land was represented by which codes. Each code would have to represent quite large areas of land, and, in general, there would be no ready made system of numbers for individual land parcels to supplement the group code.

It is further hoped that, ultimately, machines will be designed which can read hand-writing directly, but this is fraught with difficulties of different writing materials, different letter styles and sizes, and the codes being written in different parts of variously sized envelopes. In the present scheme, a postal clerk copies the code by typing it onto the envelope in a series of phosphorescent dots which the sorting machine can discern.

The Post Office do not envisage doing away with conventional "house-in-street" addresses in favour of the postal code which can so easily be incorrectly written, but rather they see the two forms of address being used in all letters.

A Note on the Central Register of Businesses

The Government Statistical Service is undertaking the task of creating a common register of business establishments and other reporting and accounting units. This will be used by all Government Departments in their statistical enquiries, and will be the basis of a data bank for analytical and information purposes. Although the system envisaged is basically one for business statistics, it may eventually be extended to include social statistics.

The core of the system now being developed is an index or register of businesses, giving the names and addresses of enterprises, companies and establishments, nature of activity (by MLH) and geographical location (by employment exchange and possibly with a twelve-digit grid co-ordinate reference). This register will include all businesses engaged in manufacturing, distribution and services, and will be used as a sampling frame for a wide range of enquiries, formerly based on registers developed by separate Government Departments. Eventually, the register will also provide the frame for the development of a data bank stored on discs.

The raw data from the enquiries relating to the register (in particular, the new production and employment surveys) will accumulate in the form of magnetic tapes, which will always include a minimum location code relating to employment exchange areas. Data about individual businesses will be highly confidential, and for most purposes aggregation to employment exchange level will be necessary; at this level it is anticipated that large volumes of summary information could be stored on disc and made available to industry and commerce, as well as to Government Departments.

(For more detailed discussion of the register and its development, content and use, see Statistical News, 1.7 and 4.5.)

British Rail Data

Permanent way:

Each divisional office maintains detailed plans of track and facilities. Headquarters staff are in the process of devising a computerised system to describe and record permanent way features. Each section of track will have a code number (similar to graph theory edges and vertices) and features and facilities will be described by a series of attribute codes (e.g. level crossing, bridges, signals, etc.). Processing will probably be on the BR ICL machine at Crewe. (BR have several machines of different makes at their divisional offices.)

Freight:

1 Consignment note analysis: This covers freight on freight trains, but excludes parcels. There is a monthly analysis of movements (tonnages, commodities) between primary marshalling yards. These are held on magnetic tape at a divisional office: data are confidential.

2 Wagon load test: In 1964, various weeks in the year were used, depending on the availability of data, to give more detail (commodity, tonnage, origin/destination, revenue, costs incurred) for freight movements. The annual totals were corrected, as far as possible, to agree with the Board's carryings for 1964. The data are stored on cards and tape, and are confidential. In 1966, a similar analysis was done for a test week in March.

Parcels:

Analysis of movements between parcels depots. The survey was confined to a 10% sample of 243 forwarding points, which accounted for 80% of the traffic. Volumes were measured in a great variety of units (e.g. gallons, mail-bags, etc.) according to commodity.

Passengers:

1 Inter-city indices: Sampling of ticket sales for major long-distance trains. Monthly intervals, 1st and 2nd class. Changes during last three years shown in index form.

2 Passenger train survey: Autumn 1968. Monthly basis. Counts and interviews with passengers for main routes. Manual recording. Files held at Marylebone Road. Similar survey in 1964, but on corridor trains only.

1. Subject **Surface Water**

Content: River flow and rainfall for all stations for which complete records exist.

Geographical coverage: Great Britain

2. Method of Data Collection

Data from recording machines are processed on the ICL 1901 computer at the Water Resources Board. Monthly diaries of daily levels, or actual tape from recording machines, are analysed by computer to produce monthly figures of mean, maximum and minimum flows. Daily values are not published.

3. Geographical Resolution

All data are for individual recording stations.

4. Statistical and Legal Restrictions

Confidentiality: None

5. Boundaries

Data refer to individual stations.

6. Forms of Data Storage

Original data: Lists of daily flows, or punched tapes of water levels from recording machines; a total of 530 stations are used in England and Wales, of which 280 supply punched tape records.

Processed data: Data on 16-channel punched tape are converted to 5-channel paper tape for computer analysis. Each river authority receives print-out copy. Daily flows are also held on 8-channel paper tape and on magnetic tape.

7. Operational Details

Originating body: Water Resources Board and Scottish Development Department, in conjunction with river authorities.

Location of supervisory staff: Water Resources Board,
Reading Bridge House,
Reading.

Frequency and dates of survey: Recording on a continuous basis by automatic machines.

Location of original data: Water Resources Board, Reading.

Location of processed data: Water Resources Board, Reading.

Publications: The "Surface Water Year Book of Great Britain" (including supplementary data provided by the Scottish Development Department) gives data for each of the stations used. Each yearly section gives monthly figures for each station, and the quinquennial supplement describes the characteristics and location of each station.

1. Subject — **Landscape Assessment (EACC)**

Content:

An exercise has been performed on the Ordnance Survey 1" maps covering East Anglia. Landscape evaluation is clearly an essentially subjective field, but it has been assumed that certain distinctive elements contributing to "landscape" are marked on 1" maps and can be plotted. The basis on which the work was done was kilometre squares which are marked on the maps, and so, by placing tracing paper over the map, the search procedure can be performed in a straightforward, if lengthy, manner.

The categories which were thought relevant for this exercise, were woodland, slope (indicated by two or more different contours in any particular square), vertical features (e.g. church towers, spires, windmills, lighthouses, monuments etc. – but it did not include those features thought to detract from the landscape, e.g. pylons), heathland, sand above the high water mark, dunes, and water features. The various features were evaluated on a point system, and the squares were coloured according to the total number of points scored.

Geographical coverage: East Anglia (as defined for the Economic Planning Region).

2. Method of Data Collection

Information derived from 1" OS maps is marked on tracing paper overlay showing which features are present: summary information (i.e. scores for each square) is presented at a reduced scale.

3. Geographical Resolution

All data relate to 1-km. squares.

4. Statistical and Legal Restrictions

Confidentiality: None (derived from published material).

5. Boundaries

East Anglia Economic Planning Region (HMSO)

6. Forms of Data Storage

Original data: 1" OS topographic maps, plus overlays.

Processed data: The information deduced from the 1" maps has been reproduced on a reduced scale to cover East Anglia.

7. Operational Details

Originating body: East Anglia Consultative Committee, Huntingdon.

Location of supervisory staff: Huntingdon.

Frequency and dates of survey: Conducted 1968/9.

Location of original data: Huntingdon.

Location of processed data: Huntingdon.

Publications: A map of landscape features, by km. square, appears in "East Anglia: a Regional Appraisal", published February 1969.

NB Work is also being done by the East Anglia Consultative Committee to try to test reactions to different features or elements of the landscape. A series of sketches has been prepared, in which features have been added to and subtracted from a variety of different scenes, in order to try and test the effect of each on the "beholder" of the landscape. This work is in its early stages, and experiments have only been tried on a small section of people, who rated the sketches on a one to ten scale. This may be compared with Fines' work in Sussex, described in "Regional Studies", Volume 2 – No. 1.

Education Statistics: an Example of a Non-planning Topic

This note describes the main range of school statistics which a Local Education Authority handles for its own purposes and for returns to the Department of Education and Science. It serves as an example of the statistical procedures and analyses in use in a non-planning field, and to show what contribution a computerised mapping information system may be able to make for non-planning purposes.

School statistics appear to be essentially hierarchical in structure, details of individual pupils being held at school level, details of classes and schools held at an Education Authority level, and details of Education Authorities held at a Central Government level. There are, of course, exceptions to this generalisation, particularly where school leavers or pupil transfer is involved.

A wide range of enquiries is undertaken on behalf of the Department of Education and Science. Normally the DES requests the Education Authority for information, and the statistics section of the authority sends a circular to the appropriate schools, asking them to complete a return which is then used to compile a summary table for the LEA as a whole. Surveys and returns for the DES include:

1 January return:	Age and sex of pupils, size of classes, numbers of teachers, numbers of immigrant pupils, by type of school (this is the only return which the DES has from each school).
2 Meals and milk:	Numbers using meal and milk services, whether free or for payment.
3 School-leavers:	Numbers, date of birth.
4 Examinations:	Subjects, numbers of candidates, results. For a sample of leavers, destination; if university, subject to be studied.
5 Married teachers:	Details of children in nursery classes, in the cases where the child's mother is a married teacher.

Addresses of pupils are normally held at the school only, and the geographical resolution of most of these data is, therefore, either the school or the education authority. (However, the School Enquiry Officer may have the addresses of children, as he has to ensure their regular attendance at school.) School catchment areas may be clearly defined, authority-wise, depending on type of school and education authority policy.

Estimation of school and equipment needs in Norwich is based on data from several sources. Heads of schools provide demand estimates based on registration of future pupils by parents. The Medical Officer provides details of numbers of children born in Norwich to Norwich parents on a year by year basis. The planning department gives advance information on the quantity and distribution of new housing, and the Building Inspectors give notices of completion for new residential developments. The population census of the GRO is not particularly useful, because of the long intercensal periods, and the Education Department relies on a special house-to-house enquiry to find out the numbers and ages of children.

The key information is the number of new pupils commencing school for the first time, each year. The house-to-house census is valid for up to four years, the figures being modified by estimates for wastage, private sector education and migration. Junior school returns indicate the numbers in the cohorts passing on to the secondary schools. The situation in Norwich is relatively simple at present because there is a fairly stable total population within the City boundary. Most suburban expansion is taking place in Norfolk. However, there are marked local fluctuations, for the new housing estates show a very distinctive family cycle, depending on the age of the houses.

Education statistics tend to be arranged hierarchically, both in terms of collection and storage. Major returns and analysis (e.g. new entrants, school leavers), take place usually on an annual basis, and the precise format of returns may vary from one authority to another. In general, the statistics are much simpler than those used for planning purposes, involving more explicit data sources, less comparison and correlation, short-term rather than long-term projection, and less geographical analysis. In a sense, the "education planning" system represents a sub-set of the "urban planning" system, although special topics, such as school meals, married teachers, etc., are more for internal administration than forward planning. In large authorities, there may be potential benefits from the application of computer techniques to education statistics, and a certain amount of pay-off from having access to a wider range of data. Applications for computer mapping would be very specialised, and might be of use in catchment area analysis.

Data Availability in Local Authority Planning Departments

Requirement: Most of the items included in the Data Availability Catalogue are derived from central government sources. If a computerised information system is to be of use to local planning authorities, it must contain not only a basic range of central government data, but also data from local sources. Part of the effort on the Data Availability Catalogue has, therefore, been directed towards establishing the type and range of data from local sources that a local planning authority might wish to include in a computerised information system. These investigations have been based upon several approaches.

Approaches:

(i) Formal approaches to an urban (Norwich) and a rural (Norfolk) planning authority in East Anglia to establish the types and sources of data that have been used in the two planning departments during a two-year period.

(ii) Extensive discussions with the East Anglia Consultative Committee to establish the types and sources of data in use in the other six planning authorities in the region. These discussions have focussed upon the degree to which Norwich and Norfolk, for which we have detailed information, are representative of the remaining planning authorities, and the extent to which information collected by more than one planning authority is gathered on a uniform and comparable basis (e.g. are the same definitions and codes used?).

(iii) Discussions with a number of planning authorities (West Riding County Council, Liverpool City Council, Hertfordshire County Council, East Sussex County Council, Greater London Council and Southwark Borough Council) and other interested bodies (West Midlands Regional Study Group, Ministry of Housing Study Group on data needs for planning, Urban Systems Research Unit at Reading University, Landscape Research Group, Countryside Commission, Centre for Environmental Studies, Llewellyn Davies and partners, et al.) covering the same range of topics.

(iv) Examination of a number of planning studies, development plans, and research reports, and publications such as the new Town and Country Planning Act, Countryside Act, and the report of the Planning Advisory Group on "The Future of Development Plans".

It is recognised that this effort is only partially adequate, and that a depth study of information needs in a planning department would require a much greater level of effort.

Heterogeneity:

All the results from these investigations, however, point to a very great variety in the information gathered and used by local planning authorities. Authorities vary considerably in size, character, expertise, and political objectives, and the net result is that, apart from data collected on behalf of the central government, described in the Data Availability Catalogue, there is very little information collected on a common basis by local planning authorities. Each authority varies its data collection effort to meet its own planning problems and resources, and this makes aggregation of data collected by more than one authority virtually impossible. The major exception to this is in the case of studies where authorities have collaborated together, such as the Norwich Transportation Study.

The attached lists show the data in use in the planning departments of Norwich City and Norfolk County Council, differentiating data collected by or on behalf of central government departments. The joint effort on the Transportation Study represents the largest extent of records collected on a common basis for local purposes. Registers of planning applications, planning decisions, tree and building preservation orders and derelict land are maintained. These form essential records, and their maintenance is partly a statutory obligation, but even these items can vary in organisation and format from one planning authority to another. Another record held by many planning authorities is the survey of existing land use. However, coverage is by no means complete, as surveys often relate to different base years, and are based upon different classifications. A Ministry of Housing and Local Government examination of the land use classifications used by eight different local planning authorities, revealed 1,500 different classification items after all possible duplications had been eliminated. *

Another important field for data collection by planning authorities is for local town centre studies. Here again, there are great differences in approach, organisation and definition, between authorities, and even between studies in different towns carried out by the same local authority.

* Quoted in a paper given by R. R. Rhodes to the Town Planning Institute Conference on the Use of Computers, Birmingham, 10th – 13th April 1969.

Problems and remedies: The heterogeneity of local data used by planning departments causes great problems of comparison, and these have been highlighted in recent years by the efforts of planning authorities to collaborate for regional and sub-regional purposes. Local government re-organisation, such as the creation of the GLC and enlarged London Boroughs in 1965, produces similar problems, and these will be multiplied if the Royal Commission on Local Government produces any changes. Efforts to standardise procedures and classifications have been made by Regional Planning Groups such as the Standing Conferences for Regional Planning in the North West, in Yorkshire and Humberside, and the South East. In addition, the Ministry of Housing has organised working groups to try to standardise procedures for land use surveys, and for the collection and submission of data under the new regulations for development plans enacted in the 1968 legislation.

Conclusion: These findings suggest that there are very few, if any, items of data collected by local planning authorities for local purposes that are collected on a common basis by a sufficiently large number of authorities to merit individual inclusion in the Data Availability Catalogue. The attached lists indicate the range and detail of data used by two reasonably typical planning authorities. Many other authorities collect and use less than this, some collect more, but nearly all lean heavily on central government sources.

In general, data collected for local purposes provide more detail and better geographical resolution for the planning authority than that available from the central goverment, although similar confidentiality restrictions may apply to members of the public. The range of basic units (persons, households, businesses, land parcels, vehicles and trips) is similar to that of central government surveys.

These findings indicate that a computerised information and mapping system for planning purposes cannot be planned with foreknowledge of all the demands that will be made of it. Any system must be sufficiently open-ended and flexible in terms of storage, coding and access, for local authorities to be able to add the data items that they wish, and perform the analyses and interrogations that they require.

Main Data Items used in Local Planning Departments: the Cases of Norwich City and Norfolk County Councils

(based on letters and documents received in November and December 1968, and on subsequent discussions. The assistance of the appropriate Planning Officers is gratefully acknowledged. (CG) indicates data collected by or on behalf of a central government department)

Subject	Norfolk County Council	Norwich City Council
Population	Census publications for 1951, 1961, 1966, plus specialised output prepared by GRO. (CG)	Ditto (CG)
	Electoral rolls.	Ditto
Housing	Local Housing Statistics, prepared by MHLG and published by HMSO (CG)	Ditto (CG)
	House-building figures, by parish, supplied by RDC's.	Slum clearance programmes; local studies for improvement areas.
		Building completions.
Employment	Employment estimates for exchange areas, by MLH, prepared by DEP. (CG)	Ditto (CG)
	Register of employers of five or more persons, prepared by DEP. (CG)	Register of unemployed persons, prepared by DEP. (CG)
	Industry, Occupation and Workplace and Residence tables for 1951 and 1961, and Ward Library for 1966, prepared by GRO. (CG)	Workplace, Occupation and Residence tables. (CG)
	Agricultural employment by parish prepared by MAFF. (CG)	

Subject	Norfolk County Council	Norwich City Council
Land use	Apart from Transportation Study material, no comprehensive and regular land use record is maintained. Local land use surveys are carried out, their extent, timing and detail being varied to meet needs.	Periodical surveys. Covered by Norwich Transportation Study material (see below under section on Traffic).
Traffic	Traffic flows at Standard Census (summer: vehicles per sixteen-hour day. A few winter counts). Prepared by Highways Department on behalf of Ministry of Transport. (CG)	Two permanent and four temporary sites. (Partly CG)
	Localised studies of traffic and parking conditions to deal with particular problems.	Ditto
	Both Authorities are collaborating on:	
	Land use/transportation study for Norwich area. This includes land use, floor space, population, employment (by zone), selected screen line and cordon counts of vehicles, plus home interviews.	
Shopping	Census of Distribution 1950 (Vols. 1 and 2). Census of Distribution 1961 (Parts 1, 2, 6 and 14) plus additional tabulations, prepared by the Board of Trade. (CG)	Ditto (CG)
	Localised studies in connection with town centre studies	Ditto
Planning applications	Register of planning applications and decisions.	Ditto
	Register of applications for industrial development.	Statutory register of applications for advertisements.

Subject	Norfolk County Council	Norwich City Council
Land, buildings and forestry	Register of derelict land (on behalf of MHLG). (CG)	Register of listed buildings. (CG)
	Register of mineral workings.	
	Register of tree preservation orders.	Ditto
	Details of roadside planting schemes.	Schedule of ancient monuments prepared by MPBW. (CG)
	Felling licences.	Register of building preservation orders.
	Managed woodland.	Corporation ownership records.
		Industrial workers' sports facilities.
Air photography	Air photographs are used in minerals, caravan, coast and recreation studies.	
Utilities and services	Comprehensive and regular records are not maintained: both Norfolk and Norwich obtain up-to-date information from the authorities on an ad hoc basis as required.	

1. Subject **Register of Electors**

Content: A list of names and addresses of those people entitled to vote in elections, both local and parliamentary. Some entries are preceded by a date which indicates that they are entitled to vote only on and after that date, having then reached their eighteenth birthday. Service voters are similarly identified by an "S" in the register. They, in common with Crown Servants and British Council staff employed outside the UK (and their spouses if living abroad with them), may be registered in respect of the address in the UK where they would otherwise reside. People resident at more than one address are entitled to be registered in respect of each. They may vote in each at local government elections, provided that they vote once only at any one local government election. Similarly, they may vote once only in a parliamentary ("general") election. There are severe penalties for infringement. Foreigners are not eligible for inclusion in the register (see 4. below).

Note: Citizens of Commonwealth countries are British subjects, and are therefore eligible (if otherwise qualified) to be registered.

Geographical coverage: United Kingdom

2. Method of Data Collection

The register is prepared by canvassers who conduct a house-to-house survey using the old register as the basis of their enquiries. They leave a form for the occupant to fill in and return to the registration officer. Large blocks of flats often receive this form by post, and, if no reply is received, are given a second chance by being sent a duplicate. If this, too, is not returned, the names of the occupants may be removed from the register, unless the electoral registration officer has reason to believe that they are still resident there.

3. Geographical Resolution

Basic information unit: Individuals listed by property where they reside.

Published data: Property-level data.

4. Statistical and Legal Restrictions

Statistical limitations: The register forms a convenient, accessible guide to population totals, but will lead to under-counting due to (a) children, not yet old enought to be registered, (b) foreigners and (c) people moving into the area after the qualifying date for the register, and who are therefore not included in the register for the area. Over-counting may also arise in respect of those who have moved away from the area after the qualifying date.

5. Boundaries

Maps produced by HMSO show parliamentary constituency boundaries. These have been at ½" scale in the past, but are now at 1 : 100,000. Ward boundaries are marked on OS 1 : 1,250 maps.

6. Forms of Data Storage

Original data: Enquiry forms left by canvassers for occupant to complete, or sent by post to occupants of flats.

Processed data: Street index of electors where practicable, otherwise alphabetical list of electors.

7. Operational Details

Originating body: Electoral Registration Officer who is also Clerk of the County Council, or Clerk of the Borough Council.

Location of supervisory staff: Electoral Registration Officer, County Council Office, or Town Hall.

Frequency and dates of survey: Annual. Comes into force on 16th February, and is extant until 15th February of the following year. Based on a qualifying date of 10th October.

Location of processed data: County Council offices or Town Halls.

Publications: The register is itself a published document, to which the public have right of access. Parts of the register may be purchased from the Electoral Registration Officer for the area, the cost being proportional to the number of electors involved.

Ministry of Housing and Local Government – Map Library

The map library is housed on the fourth floor of the Whitehall building of the Ministry of Housing and Local Government (MHLG). Slides and pamphlets are included in the library, but the bulk of the material is a large collection of maps classified and arranged by subject and scale. There are also references for locality, reports, etc. There is a very great variety of maps, at many scales, and using many different kinds of symbols. Both mass-produced printed maps and unique manuscripts are included.

The library is particularly valuable as a source of boundary information, especially as described below:

(a) Local authorities and parishes: For counties in England and Wales, administrative diagrams held are as published by the Ordnance Survey from about 1900. Until recently these were at ½" = 1 mile scale, but the English county diagrams have been changed to 1 : 100,000 scale with the National Grid lines shown at 1 km. intervals, whereas post-war editions of the ½" series had a 10 km. interval. The admin-sitrative information for Scotland is on the outline version of the standard 1 : 250,000 (about ¼" = 1 mile) series sheets.

The boundaries and names are in red on the topographic base (in grey) with colour tints for the urban authority areas. Parliamentary constituency areas are shown in blue (formerly green). New editions are published periodically, and the current edition annotated by the Map Library for changes announced between editions. All amendments to local authorities are effective from 1st April in any year, but those affecting parishes can have varying dates. Changes are made either under the Local Government Act, 1933, or that of 1958, which gave effect to certain of the Local Government Commission proposals.

A 1 : 625,000 (about ten miles to one inch) map of the Administrative Areas in Great Britain is published by Ordnance Survey, and is usually revised annually. This is in two sheets, and is printed in colour on an outline topo-graphic background. The Ministry has a limited printing of a one-sheet map for England and Wales of the local authority names and boundaries information alone, in one colour. A twenty-mile (reduced) version is also held.

Wards – Alterations to wards are also made under the 1933 Local Government Act, but are notified to MHLG by the Home Office. They are comprehensive-ly mapped by the General Register Office and by MHLG, who have a unique series of maps in the Map Library. This series has local authority, parish and ward names and boundaries (in red) on an OS outline topographic base (grey) at 1 : 25,000 scale. A complete set of sheets for England and Wales has all information as at Census date 1961 (23rd April 1961); areas for which any change was made up to April 1966 (10% sample census date) are mapped on a supplementary sheet, and alterations since that date up to the next census (April 1971) are being added to a further supplementary sheet. A ten-mile map is annotated to act as a key to these.

There is also a set of 1 inch = 1 mile (coloured) maps with local authority, parish and ward names and boundaries for 1961, and a series is being drawn now for 1966. This latter will have all names and boundaries (in black) on a translucent medium for overlaying to a one-inch map.

Enumeration districts — Photocopies of the GRO 6″ = 1 mile maps of the boundaries and code numbers for England and Wales as at the 1966 (sample) census date are held for reference, and microfilms for further reproduction.

(b) Board of Trade: Census of Distribution: Shopping centres, as defined by the Board of Trade for the 1961 Census of Distribution, have been plotted by streets by MHLG, mainly at 25″ scale.

(c) Employment exchanges: These Local Office areas of DEP (employment exchanges plus component sub-offices and branch offices) have been drawn on translucent overlays to the 1 : 100,000 scale administrative diagrams for England, and direct on to copies of ½″ county diagrams for Wales. Changes are recorded as notified by DEP. Maps are produced annually at the 10-mile scale, showing the exchange boundaries as at 30th June.

Other Reference Maps

Of special interest among the reference maps are:

(a) A 1″ set (on OS topographic base) showing the location of Forestry Commission Lands (distinguishing those owned and those protected by Deed of Dedication), Tree Preservation Orders (to February 1969), National Trust Lands, the Crown and the Duchies of Cornwall and Lancaster Lands;

(b) A series at 2½″ = 1 mile, showing Nature Reserves, Sites of Special Scientific Interest, National Parks, Areas of Outstanding Natural Beauty;

(c) ¼″ = 1 mile maps of proposed and approved Green Belts (1″ = 1 mile for the Metropolitan Green Belt);

(d) Ancient Monuments series (at 2½″ = 1 mile) of sites listed by the Ministry of Public Building and Works;

(e) 6″ = 1 mile maps of approved country parks and picnic sites, also of Conservation Areas under the Civic Amenities Act, 1967 (some are on a larger scale);

(f) Large-scale maps of various shopping centres, showing individual shops by type.

All the above [(a) to (f)] are overdrawn or printed on to an Ordnance Survey base map. There is also a set of 1″ maps locating Ministry of Defence Lands; these maps are usually available for reference, but access must first be authorised by the Ministry of Defence. This set is at a small scale, mostly based upon census material.

The MHLG Library is not a public map library, such as those at the RGS and British Museum, and is not normally available for public use of published maps. There is, however, a large number of unpublished maps, and these are normally made available to bona fide users for reference.

Planning Information in Nature Conservancy Map Library

The Nature Conservancy Map Library is housed in the basement of the Belgrave Square headquarters building. The Conservancy has responsibilities throughout the United Kingdom, but there are separate map facilities to serve Scottish needs.

The map collection is built round a master index of cards referring to individual sites, filed alphabetically by the name of the site. The index includes Sites of Special Scientific Interest (over 2,000 of them, defined under Section 23 of the 1949 National Parks and Access to the Countryside Act), National Nature Reserves (128 of them, defined under Section 16 of the same Act), and Local Nature Reserves (defined by County Councils under Section 21 of the same Act). There are also Forest Nature Reserves (by arrangement with the Forestry Commission) and two Nature Reserves (by arrangement with the Duchy of Cornwall).

The card index is a key to the map collection, which is arranged basically by county. Maps of individual sites are always superimposed on Ordnance Survey base maps, ranging from 1 : 2,500 scale to 1" scale, depending on the size of the site and the scale availability. There may also be field surveys prepared by Nature Conservancy staff, usually not on OS base maps.

County key maps (covering all categories of site) are in course of preparation. Sites are shown by a spot reference, except those over 1,000 acres in extent, for which a boundary is shown. At present, twenty-eight counties have been completed (in England and Wales), mainly in the upland areas. A small-scale map of the United Kingdom is available showing location of NNR's, LNR's and FNR's in diagrammatic form.

1. Subject: **Second Land Utilisation Survey**

Content: Field work done on 6" base maps of the OS (cf. Dudley Stamp's work) and published at 2½" (1 : 25,000 scale). The maps depict 64 categories of land use, overprinted in 11 colours. Maps are designed to be read at two levels: the first impression is one of colour differences, denoting main groups of uses; at a more detailed level, variations of tone and other subdued cartographic devices indicate subdivisions within each group.

Derived maps: (i) A wildscape atlas is being prepared to show the different vegetation categories in colour, with eight different overlay patterns to indicate habitats. The atlas will consist of 162 pages, with maps at a scale of 1 : 100,000.

(ii) The concept of land use territories is being used as the basis for another set of maps. Five types of territory (townscape, farmscape, wildscape, urban and marginal) will be mapped after careful analysis of the 2½" land use maps. Any of the land use categories can appear in any of the territories, but in three of the latter, (townscape, farmscape and wildscape), certain categories will be dominant, whilst in the fringe areas of conflicting interest (rurban and marginal), there will be co-dominant categories occurring. The following conceptual diagram illustrates the relationships between the categories:

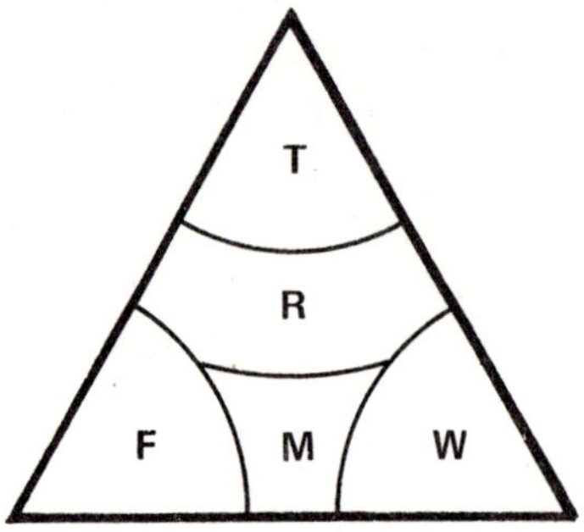

LEGEND
T – Townscape
R – Rurban
F – Farmscape
W – Wildscape
M – Marginal

The detailed production techniques of these maps have yet to be decided upon, but it is envisaged that they will finally appear at a scale of 1 : 400,000, which will be a x 4 linear (or x 16 areal) reduction from the scale at which the material is being drawn up at present.

Miss Coleman has the idea of using a machine which she calls a "morphogram keyboard" (which does not, as yet, exist) to record the land use territory details onto magnetic tape. It would then be possible to analyse the material in many different ways, and the input technique would, she claims, be a great deal quicker than "conventional" digitising.

Geographical coverage: Great Britain

2. Method of Data Collection

Field sheets are assembled in London from the various surveyors.

3. Geographical Resolution

Individual crops and animals are identified in more detail on the field sheets than appears on the published maps.

4. Statistical and Legal Restrictions

None.

5. Boundaries

The maps themselves are the boundary source.

6. Forms of Data Storage

Original data: 6" OS base maps, (843 for England and Wales).

Processed data: Sheet lines of standard sheet are 20 kms. x 10 kms., with the exception of some coastal sheets. Each sheet represents 78 square miles of terrain.

7. Operational Details

Originating body: Organised by Dr. Alice Coleman of King's College, London, in conjunction with the Geographical Association

Location of supervisory staff: King's College, London.

Frequency and dates of survey: In England and Wales, the survey was carried out between 1960 and 1969. In 1964 it was extended to Scotland. (This survey was a natural development from the first land utilisation survey, organised by Dudley Stamp in the 1930's.)

Location of original data: Manuscript maps are held in King's College: these cover the whole of England and about half of Wales. 25% of Scotland is also available, some of the material being taken from the 1950's Crofting Survey. The Scottish part of the survey is organised by a separate sub-committee.

Location of processed data: Printed maps are available from King's College, and from Stanfords in Long Acre, London.

Publications: 112 sheets have been published to date (June 1970). Some of these sheets are of places of special interest, whilst others have been published to form series of adjoining sheets, e.g. South Wales, London, East Kent, York/Leeds/Doncaster, coast of Northumberland/Durham.

Ministry of Technology – Production Statistics

The Ministry of Technology collects statistical data from companies engaged in the manufacturing industries.

The response to the statistical enquiries is generally on a voluntary basis rather than statutory. The enquiries are of two principal types:

(a) In the mechanical, instrument and electrical engineering sectors, enquiries are made for the value of deliveries and orders, total and for export, to all companies employing twenty-five or more people. Enquiries are made monthly to the larger companies, and quarterly to the others – more than 4,000 in total. Currently, there is a variety of reporting unit, ranging from establishment to enterprise. The main purposes of this enquiry are to produce leading indicators, such as the engineering element of the index of Industrial Production; the Index of Engineering Orders and Deliveries; values of deliveries of certain mechanical, instrument and electrical engineering products. Summaries of these data are published in the Board of Trade Journal, the Monthly Digest of Statistics and the Annual Abstract of Statistics. (The figures published monthly are calculated to take account of estimates of the activities of companies included in the quarterly enquiry.)

(b) Enquiries to companies for details of deliveries (or of production, sales etc.), within particular product/industry sectors of the range of the manufacturing industries. There are more than seventy such enquiries. Each enquiry aims to cover at least all the major manufacturers in a sector. The results of the enquiries appear variously in such publications as the Business Monitor Series, the Monthly Digest of Statistics, the Annual Abstract of Statistics and the Board of Trade Journal.

The response to these enquiries comes from a variety of reporting units. This places severe limitations on any regional analysis of the data, as returns frequently cover activity in establishments in widely different geographical locations. Although the enquiries are non-statutory, disclosure is restricted within the terms of the Statistics of Trade Act, 1947.

When the Business Statistics Office is in full operation, these surveys will be recast, and extended to form part of a comprehensive range of production enquiries (see paragraph 2 (c) of the section of the Appendix relating to the Census of Production). The reporting units used for these enquiries will, as far as possible, be the separate geographical location, but returns covering combinations of establishments in different areas will continue to some extent. The scope for producing separate figures for different areas will continue to be constrained by the need to avoid disclosing the activities of individual firms.

Home Office Statistics – a General Note

The Home Office compiles statistics relating to each of its major areas of responsibility, and to a number of miscellaneous control functions which it exercises.

Since the installation of the first computer in 1963, some changes have been made in the collection and processing methods and scope of data. A large proportion of the statistics is processed by computers from punch card or paper tape input, and in some fields it has now become possible to produce more detailed tabulations than before.

Criminal and penal statistics comprise the largest area of the Department's statistical work. At present these are kept separately at three stages: offences known to the police, court proceedings and their results, and penal treatments. The Report of the Departmental Committee on Criminal Statistics (HMSO 1967) recommended that all such records should be linked within a single computer organisation, so that it would be possible to trace the career of an individual, or to follow up the consequence of an offence from the time it was recorded by the police.

These proposals, if adopted, would extend the possibilities for statistical analysis and research, but it must not be assumed that the full data available, particularly that referring to individual criminal histories, would be suitable for a general data bank.

Selective List of Home Office Statistics

(a) Criminal Statistics:

Offences	Proceedings	Penal Treatment
Motor vehicle offences	Higher Court proceedings	Prisons
Offences involving firearms	Appeals	Borstals
Homicide	Prosecutions by the Director of Public Prosecutions	Detention centres
Offences of Drunkenness		etc.
Dangerous Drug offences		Receptions
Speeding offences		Discharges
etc.		Population
		Offences and punishments

(b) Child care and juvenile delinquency:

(i) Children in Care	(ii) Work of Children's Departments	(iii) Offences committed by Juveniles
Voluntary homes		Proceedings
Local authority homes		Treatment

(c) Police: Strength of forces:

Annual returns by police forces.

(d) Miscellaneous: e.g. Foreigners entering the UK
Experiments on living animals
Betting licence permits
General election expenses
Liquor licences granted.

Ministry of Public Building and Works: Statistics

The statistics of the construction industries and of the building material industries, collected by the Ministry, provide the economic background data required for estimating the current situation in the industries, and for estimating future trends.

There are basically three main areas of enquiry (see below): (a) production, (b) employment and (c) short-term demand for the industry. The most important surveys are carried out quarterly, and most of these data are at present transferred to 1" magnetic tape for processing on ICT 1300 series computers at Hastings (conversion to ½" magnetic tape for processing on ICL 1900 series computer is shortly to be undertaken).

Of particular importance for regional planning purposes are the statistics of new orders and contracts for construction, which are analysed by region (and with less statistical reliability, for local authority areas). These provide an estimate of the amount of work in the pipeline, and are a guide to the future load and its distribution.

Spatial analysis of these data is restricted by (a) confidentiality, (b) the limited geographical resolution of the data (normally Economic Planning Regions are the smallest areas for which data are presented), (c) statistical reliability (this applies only to the survey of new orders and contracts) and (d) location codes often relate to the place of registration of firm or public authority, and not to the actual place of construction and employment (see below).

MPBW Statistics

Production:

Building materials	Construction
Bricks	Figures relate to work carried out by firms and public authorities registered and located in the Regions, and not to location of work.
Cement	Value of work by type – by private builders
Sand	Value of work by type – by public authorities.
Gravel	

Employment (Construction): Operatives employed by type of work – Private employers.

Operatives employed by type of work – Public employers.

Administrative, professional, technical and clerical employees employed.

NB Employment figures relate to firms and public authorities located and registered in the Regions, and not to the location at which the employees work.

New Orders and Contracts: The sample survey relates work to the area in which the construction is to take place, by type of work and value of contracts.

Appendix 2

A standard format for the exchange of cartographic data

Proposed by T. Lang, PhD.

It is commonly agreed that the digitising of cartographic data is a costly and time-consuming practice. It would be a great waste of valuable resources, therefore, to digitise the same information more than once. If this is to be avoided, there must be some straightforward means for the exchange of data between different organisations. This note attempts to define a standard format for data exchange.

The requirement to define a standard medium for the exchange of data does not mean that this data must be stored in a common format within all computing systems. The study of automated cartography has not progressed sufficiently to make it possible to recommend a standard design of data bank, indeed this may never be possible. The approach has been rather that information can be standardised just after the stage where it has been digitised, checked, and if necessary corrected, but before the point where it is assimilated into a data bank.

For this reason, it is anticipated that co-ordinate data exchanged by these means will usually be specified as the table co-ordinates taken direct from the digitising process. It will be up to the installation receiving the data to apply the necessary transformation into the units which it requires for assimilation into its own data bank.

The transmission of data in the form directly as taken from the digitiser has the advantage that it should be possible to recover the original map exactly (to the accuracy of the digitisation process). However there is no reason why the installation issuing the data should not apply any arbitrary transformation which it feels is useful – e.g. a transformation into latitude and longitude (with a suitable change of scale and origin to enable the resultant numbers still to be expressed in the standard format). In the case of the Experimental Cartography Unit, the original table co-ordinates will be merely rotated and translated so that the map described has its axes parallel to the numerical map axes, and with the origin at some nominated point (usually the south west corner).

1. Physical tape specification

1.1 7-track 'IBM-compatible' magnetic tape: Punched cards and paper tape would be too bulky. Most computer installations, even non-IBM installations, have some provision to use 'IBM-compatible' tape. Newer installations are generally biased towards 9-track tape, but even these make some provision for handling 7-track tape.

1.2 Binary-coded-decimal character recording: Straight binary recording would make for more efficient use of tape. However, the use of character recording has the advantages:-

(a) Data may be accessed directly by standard FORTRAN programs. (Basic machine code programs may be developed where local circumstances make this necessary.) The FORTRAN character code is given in the note on Tape Character Code at the end of this Appendix.

(b) In the event of uncertainty, it is always possible to list the tape on a line printer, and decipher the results visually.

It is felt that the use of BCD characters, plus the normal hardware parity checking make it unnecessary to add further error protection (e.g. inserting of checksums).

1.3 Even parity: This is normal practice where BCD recording is used.

1.4 Recording density 556 characters/inch: Alternatives are 200 or 800 characters/inch. The higher the recording density, the more efficient the use of tape, but the more susceptible the recording process becomes to hardware errors. 556 characters/inch is the best compromise. It is recommended that installations which have a choice of recording densities should make provision to use the alternatives if necessary.

1.5 Record length 120 characters: For simplicity it is suggested that all records should be of exactly 120 characters. The longer the record length, the more efficient the use of tape. However, for diagnostic purposes it is important to be able to list the data on a line printer. A common carriage width for printers is 120 characters, though some printers allow up to 160 characters.

2. Record format specification

Three different record formats have been defined:-

2.1 Header information: The first five records describing a map comprise information specifying general features of the cartographic data which follows – e.g. the area covered, the scale, the source from which the material was digitised, etc. This information can be in free format (i.e. accessed by FORTRAN via the 'A' or character format), and hence may only be interpreted by printing it and deciphering it visually. It has not been felt possible to define the record format any more precisely since the requirements are likely to vary from case to case.

2.2 Point information: These records give information related to point, as opposed to continuous line, information.

Character		Information
0 – 1		Mark number – denotes the type of information described by this record. Mark numbers so far assigned are:
	3	Start of line. A start of line record must be followed by records describing the course of the line (section 2.3).
	4	Individual point.
	5	Cornerpoint. Practical experience has shown it important that a map should contain a number of reference points, especially where several maps are later to be juxtapositioned. Usually these will be the four corners of a bounding rectangle.
	6	End of map. The following record is the start of information for another map.
	30	End of tape. As for end of map, but no other maps follow. This record will be followed by a standard 'end of file' mark.
2 – 3		Correction code. This is used where the current data is intended to update earlier information. This is discussed in section 2.2.1 below.
4 – 8		x co-ordinate. The range of co-ordinates is -9999 to 99999. No attempt is made to specify the units in which the co-ordinates should be given, as these may vary from one application to another. The unit of measurement should generally be specified in the header information. In the case of the ECU, these units will usually be 0.02 mm.
9 – 13		y co-ordinate.
14 – 15		Number of feature codes which follow.
16 – 21 22 – 27 112 – 117		Maximum of 17 6-digit feature codes. The significance of each code must vary from application to application, e.g. 1st code = hieght above sea level in feet; 2nd code = annual rainfall for 1962 in inches.
118 – 119		Blanks

2.2.1 Correction codes: The correction codes permit the giving of correction commands related to specific co-ordinate positions, e.g. 'delete the line which passes through this point'. This pre-supposes that the correction co-ordinates have been transformed to the same origin and direction of axes as the original map. The correction codes so far assigned are:-

0	Not a correction (zero characters or blanks)
1	Delete line which passes through this point
2	Delete point which has these co-ordinates
3	Insert this line (must be associated with mark number 3)
4	Insert this point
7	Change line feature codes to the codes given in this record
8	Change point feature codes to the codes given in this record.

2.3 Line information: A start of line record must be followed by further records defining the course of the line. To make efficient use of the tape, the line will be specified as a string of relative co-ordinates, denoting the change Δx, Δy in x, y from the previous point. Each 120 character record will be made up of 30 pairs of Δx, Δy relative co-ordinates, each of two digits. The range of each relative co-ordinate value will be 01 to 99, with a shift of origin, so that 50 represents the zero increment of position.

e.g.

Number on tape	Relative co-ordinate increment
52	+2
51	+1
50	0
49	-1
48	-2
47	-3

Zero values for (shifted) Δx, Δy will be reserved to indicate the end of line. The remainder of the record will also be packed out with zero values.

3. Recommendations

1 The above format should be accepted as the standard for the exchange of digitised cartographic information.

2 There should be regular consultations between interested organisations for the extension of the standard in the light of developments in digitisation practice.

Tape Character Code

Number on Tape	Number in Store	Character	Number on Tape	Number in Store	Character
00			40	40	–
01	01	1	41	41	J
02	02	2	42	42	K
03	03	3	43	43	L
04	04	4	44	44	M
05	05	5	45	45	N
06	06	6	46	46	O
07	07	7	47	47	P
10	10	8	50	50	Q
11	11	9	51	51	R
12	00	0	52		
13	13	=	53	53	$
14	14	'	54	54	*
15			55		
16			56		
17			57		
20	60	space	60	20	+
21	61	/	61	21	A
22	62	S	62	22	B
23	63	T	63	23	C
24	64	U	64	24	D
25	65	V	65	25	E
26	66	W	66	26	F
27	67	X	67	27	G
30	70	Y	70	30	H
31	71	Z	71	31	I
32			72		
33	73	,	73	33	.
34	74	(	74	34	)
35			75		
36			76		
37			77		

All numbers are in octal representation. In most systems the exchange between representation in core and on tape is performed by hardware. This character set is a subset of the original IBM 6-bit FORTRAN character set. The set has been left as large as possible, whilst omitting those characters known to cause difficulties on input in specific installations. Tape codes for which both the number in store and printing character have been left blank should not appear on the tape.